T0138925

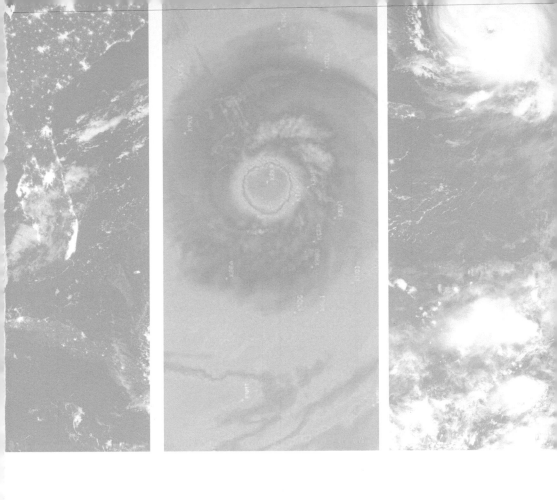

PHYSICS AND
FUTURE OF HURRICANES

PHYSICS AND
FUTURE OF HURRICANES

Edward L. Wolf

JENNY STANFORD PUBLISHING

Published by

Jenny Stanford Publishing Pte. Ltd.
101 Thomson Road
#06-01, United Square
Singapore 307591

Email: editorial@jennystanford.com
Web: www.jennystanford.com

British Library Cataloguing-in-Publication Data
A catalogue record for this book is available from the British Library.

Physics and Future of Hurricanes

ISBN 978-981-4968-54-6 (Hardcover)
ISBN 978-1-003-33125-4 (eBook)

Contents

Preface

This book is about hurricanes, prompted by a discovery that suggests they will become more powerful with global warming. It is intended to provide, at a college physics level, a basic understanding of hurricanes emphasizing the flow of energy into and out of these storms. So it is close to a textbook, covering some material that might be taught in meteorology or atmospheric physics courses. But it is centered on a new discovery that is not in any existing textook.

It turns out that hurricanes, as revealed by the new discovery, are usefully regarded as a separate phase of matter, bringing in characteristic temperature dependences near their transitions. The role of phase change in understanding hurricanes brings in the 20th century discoveries in theoretical physics relating to critical phenomena with non-intuitive values of the critical exponent β entering the formula $P = \text{const } (T - T_c)^{\beta}$, where P is a characteristic strength parameter, or order parameter, of the phase of matter appearing at T_c. In the new discovery on hurricanes, it appears that, taking the wind velocity as the order parameter P, the critical exponent is near 1/3. We find, in a second discovery, that a small correction to this value is brought in by the complicated physics of the renormalization group, which earned K. G. Wilson the Nobel Prize in Physics in 1982.

There are few books on hurricanes, none that include the new discovery. An excellent non-technical book is *The Divine Wind: The History and Science of Hurricanes*, by Kerry Emanuel, Oxford University Press, 2005. The ideas of atmospheric physics describing rainstorms are presented in that book with care and clarity. A technical summary of hurricanes was provided in the edited volume of Murnane and Liu in 2004.

The source article for the present discovery appeared only in 2005, so a great deal has changed since the date of Murnane and Liu's book. The source article of 2005 has been cited more than 3500 times, and it is surprising that not one of those citations recognized a central implication of that paper, that a strength parameter for hurricanes, the power dissipation index PDI scales as PDI = const $(T - T_c)$. That simple temperature dependence is still absent from the meteorological literature of hurricanes. If those thousands of publications were not informed of the basic law PDI = const $(T - T_c)$, implied by Emanuel in 2005, can one have confidence in the subsequent poorly informed results? This literature includes some ideas that seem non-physical, including the notion of "non-locality" that the hurricane is strongly influenced by remote conditions. A second non-physical idea is that the onset temperature of hurricanes will adjust as the climate warms to keep the strength from increasing with sea surface temperature. The T_c is plausibly determined by the properties of water, its latent heat of vaporization, and its vapor pressure, hardly subject to change with time. It appears that there is a need for an introductory physics-based book explaining hurricane properties and likely future changes. The further interest in such a book is that the new discovery of the critical temperature dependence implies a faster increase of hurricanes' destructive power with rising local ocean temperature, T.

The author has benefited from conversations with colleagues as well as from anonymous reviewers of his recent publications, "Critical behavior of tropical cyclones," *Theoretical and Applied Climatology*, 139(3), 1231–1235, (2020), and "Precise prediction of hurricane power vs ocean temperature," *International Journal of Atmospheric and Oceanic Sciences* 5(1), 1–5 (2021).

The author thanks Mr. Kunal Mehta for help in preparing the manuscript, and also Prof. John DiBartolo, Prof. Lorcan Folan, and Ms. DeShane Lyew of the Applied Physics Department at the Tandon School of Engineering for office space and secretarial help. He thanks his wife, Carol, for supporting this project.

Edward L. Wolf
October 2022
Brooklyn, New York

Chapter 1

Introduction: A Physics-One Look at Hurricanes

Hurricanes are the most energetic and destructive of natural atmospheric events. *Huracán* is the Mayan God of wind, storm, and fire, related to the Caribbean Indian God of evil, represented by a whirlwind symbol. When Spanish explorers came to the Caribbean in the 1500s, they discovered also the fearsome tropical storm known as the "Huracán." This word remains the Spanish for the hurricane. It is reported that Christopher Columbus, who sailed in 1492, brought the word back to Europe, and the word hurricane appears in the writings of Shakespeare.

The ocean waves just outside the "Eye," the quiescent circular center of the hurricane, are dangerously large, approaching 100 ft (about 30 m) in height, and have notoriously destroyed ships, including modern warships. In earlier times, two attacking armadas of Kublai Khan, up to 900 ships, sailing toward Japan, in 1247 and later in 1279 AD, were thus destroyed, prompting the Japanese to describe typhoons as "divine winds."

Such cyclonic tropical storms, known also as Typhoons and Cyclones, can be monstrous. Indeed, the cover photograph of this book shows Hurricane Irma, on September 8, 2017, center, and directly east of the Yucatan peninsula, as bigger than the state of Florida. The length of Florida, visible to the northwest of

Physics and Future of Hurricanes
Edward L. Wolf
Copyright © 2023 Jenny Stanford Publishing Pte. Ltd.
ISBN 978-981-4968-54-6 (Hardcover), 978-1-003-33125-4 (eBook)
www.jennystanford.com

central Hurricane Irma, in our cover image, is given as 447 miles or 721 km. The picture shows the hurricane as extending up to the top of the troposphere, around 12 km, but it is known separately that the high azimuthal winds occur mostly near the ocean, to altitudes on the order of one to 5 km. Looking at the image of Irma, one can see that the radius is on the order of 600 km. If you can see three huge hurricanes at once, they cannot be too rare; there are typically 80 such storms globally per year. And they can be even larger than those shown here. The biggest such storm was Typhoon Tip, which struck Japan in October 1979 with a diameter of 1380 miles (2200 km).

The landfall wind speed of Hurricane Michael in October 2018 is reported as 155 mph, which corresponds to 134.7 knots and 69 m/s. A direct dropsonde measurement from an airplane flying in Hurricane Isabel on September 13, 2003, showed (Davis and Paxton, 2005) a speed of 233 mph or 104 m/s. We take this directly measured speed as an important data point. It is a basic puzzle how such high speeds can arise spontaneously and be organized so completely into circular motion.

The energy of motion, the wind energy, in such a storm, is the integral of $1/2 \, \rho v^2$ over the volume of the storm, with v the wind speed and ρ the mass density of air, about 1.22 kg/m^3. Hurricanes occur in a wide range of sizes, we take a typical radius as 600 km (373 miles), with strong azimuthal air motion persisting up to a height of at least 5 km. (An authoritative and detailed review of hurricanes is given, e.g., by Emanuel in 1991.) A typical (peak) wind speed is suggested as 60 m/s by Emanuel (1991), in Table I on p. 191.

1.1   Some Simple Estimates

The energies of hurricanes are of central interest, and we now make simplified, rough estimates of typical power and energy. The central feature of the nearly circular hurricane is its rotation. It is approximated as a vortex with wind speed increasing linearly with radius from the center. A first-order approximation is that the whole disk rotates as if it is a solid body. This is known to be only partially correct, and a further estimate is a Rankine vortex, where the solid-body rotation stops at a specified radius R_{max} and beyond that, the speed falls off inversely as V_{max} (R_{max}/R). On the other hand, the power law outside R_{max} is $1/R^{1/2}$ according to Emanuel (1991), and the distribution is smoothed, as suggested by Fig. 1.1 (Lu et al., 2018).

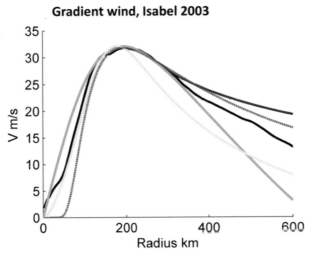

Figure 1.1 Wind profile for Hurricane Isabel of 2003, as modeled in five separate estimates by Lu et al., 2018. For the estimate in the text, the lower green curve was chosen, representing that the outer radius of the storm is 600 km. In light of the direct report of Davis and Paxton, we renormalize the vertical scale to a maximum of 104 m/s at a 200-km radius. Reproduced from Lu et al., 2018. © American Meteorological Society. Used with permission.

1.1.1   Kinetic Energy of Air Motion

We take a Physics-One look at a disk-shaped hurricane of radius R and height H, represented by Hurricane Isabel of Fig. 1.1, where we renormalize the peak speed to 104 m/s as reported (Davis and Paxton, 2005) in the direct measurement. As an approximate, back-of-the-envelope estimate, we take the average wind speed v = 52 m/s (116 mph), R = 0.6 Mm, and H = 5 km. The formula is

$$K = \pi R^2 H \, 1/2\rho v^2, \tag{1.1}$$

where K is the kinetic energy, R is the disk of radius, and H is the height. Taking the air density as 1.22 kg/m^3, this energy is 8.18×10^{18} J = 8180 PJ. This might be expressed in kilowatt-hours, kWh, a commercial unit of energy, cost about \$0.14 per unit in New York, equal to 1000 W × 3600 s = 3.6×10^6 J. Thus, our exemplary hurricane energy is 2.27×10^{12} kWh, with a value of \$318 billion. This large value is a rough estimate, and keep in mind that nature provides a wide range of hurricane sizes and strengths.

The energy of the hurricane is set by a balance of generated power vs dissipation power, the latter mainly creating dangerous waves on the sea underneath. An estimate of the hurricane's power, and its rate of energy generation, can be made from its rainfall. This follows from the assumption that power comes from the released energy, 540 cal/g = 2,259 kJ/kg from condensing water vapor to liquid, which happens in the hurricane. This energy is released initially as heat, an increased temperature, but by the mechanism of convection gets turned into large-scale kinetic energy. We think of the energy flow in the hurricane as follows:

The hurricane in its mature state takes up water vapor just above the sea surface and returns a large amount of water to the sea as rain, after losing some to evaporation from the top of the storm (Fig. 1.2). The change of water from vapor to liquid form releases 2,259 kJ/kg of energy. Since the mature storm is fully formed and stable, with its clouds, humidity distribution, temperatures, and internal kinetic energy structures unchanging, the ongoing energy release from condensation can be either as radiation from the top of the clouds, which we take as negligibly small, or as mechanical power, the release of kinetic energy that mostly creates the huge waves that sink ships, but can also lift the roofs off houses if over land. Another portion of mechanical power appears to be the flow of high-speed air out the top of the storm. It is known that thunderstorms have strong updrafts and these carry mechanical power. We will see below direct evidence that Hurricane Erin on September 12, 2001, created a huge updraft in its area extending from sea level to the top of the troposphere.

As a considered estimate of the average rainfall under hurricane Isabel, from Fig. 1.2, (Lu et al., 2018), we take 2.5 mm/h. This value, ~ 0.1 inch/h, for a large hurricane, is a spatial average, during the peak of the storm. The actual rainfall comes unevenly in localized regions (near the core, under the eyewall, and in spiral rain bands) and can locally be much larger. For example, 6 inches in 1 h was reported at one location in Houston, Texas during the peak of Hurricane Harvey, in August 2017. In that storm, the rainfall rate and the wind speed increased, in unison, just before landfall, apparently due to hovering over warmer water just off-shore.

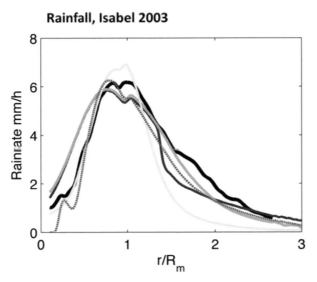

Rainfall, Isabel 2003

Figure 1.2 Rainfall modeled for Hurricane Isabel 2003. R_m represents the peak in the wind speed distribution as shown in Fig. 1.1. From this, emphasizing the yellow curve, we take the average rainfall rate as 2.5 mm/h out to a radius of 300 km, taking R_m as 200 km. Reproduced from Lu et al., 2018. © American Meteorological Society. Used with permission.

A central fact, for temperature relevant to the tropical ocean, is that the vapor pressure of water increases exponentially with temperature. The energy released per gram of water vapor, condensing to liquid, is 540 cal, which equals 2,259 J, with 4.184 J/cal. The condensation energy for water is then 2.259 MJ/kg, equal to 2.259 GJ/m^3 with a water density of

1000 kg/m^3. Per square meter, at a rainfall rate of 2.5 mm/h, corresponds to 0.0025/3600 m^3/s and thus to a kinetic power density of 1.56 kW/m^2. Thus, 2.5 mm/h of rain is accompanied by a power density of about 1.5 times that of direct sunlight. By comparison, we can almost neglect the radiation power from the top of the hurricane, where we assume the temperature is 200 K, that power is $\sigma_{sb} T^4$ = 90.7 W/m^2, only 5.6% as much. Here, σ_{sb} is the Stefan-Boltzmann constant, 5.67 × 10^{-8} W/K^4 m^2. The energy continually released by condensation into rain is, eventually, primarily released in motion of the air, $1/2\rho v^2$, corresponding to the large wind speed, in both azimuthal and vertical directions. A portion of this condensation-derived kinetic energy that is siphoned off each second, becomes the dissipation roiling the ocean.

If the hurricane is disk-shaped, with an area πR^2, taking R = 0.6 Mm, but the rain is limited to half that radius as shown in Fig. 1.2, then the total power of the hurricane is 441 TW. This is a gigantic number; for comparison, the total power use globally is about 17 TW, about 3.9% as much. On the other hand, 100 TW is given in a standard textbook for the total latent energy release in an average tropical cyclone, and it appears that Isabel is larger than average. At this power of 441 TW, the time to build up to the total stored kinetic energy of 8.18 × 10^{18} J for the Physics-One hurricane would be 5.2 h. (We are assuming height is 5 km. The actual formation of a hurricane takes much longer.) Another way of saying this is that the stored energy is about 18,500 times the energy gain per second. To stop power generation, the wind would die down in 5.2 h, with the energy mainly roiling the sea under the hurricane.

1.2   Energy Flow and Power Density

The power carried by the wind of speed v is proportional to v^3. To understand this result, consider an area A oriented perpendicular to a flow at speed v of fluid of density ρ. In one second a length $L = v$ containing mass $M = Av\rho$ will pass through the area. Thus, $dM/dt = Av\rho$. This represents a flow of kinetic energy $dK/dt = dM/dt \, v^2/2$, so that power, P, crossing area A is

$$P = dK/dt = dM/dt \, v^2/2, \text{ or}$$

$$P(A) = A\rho \, v^3/2. \tag{1.2}$$

If we assume, for our disk-shaped model of radius R and thickness H, the azimuthal speed is constant at 52 m/s through the cross-section area RH, then the wind power from this formula is 257 TW. (This appears to be an overestimate, as we will see shortly.) This can be compared to the 441 TW found above from the latent heat release in the rainfall. It is reasonable that the difference, 184 TW on this estimate, flows additionally into vertical air motion, strong updrafts, within the eyewall thunderstorm clouds.

A second estimate of the azimuthal wind power is possible considering the rotational nature of the situation.

1.3 Rotational Aspects of Air Motion

Our ballpark analysis of the energetics of the Isabel-sized hurricane so far overlooks the organized nature of the air motion. We now assume that a large part of the kinetic energy is in a circular motion, the inner part of the air mass rotates nearly like a solid body. So, a second Physics-One approximation of the hurricane is as a disk-shaped rotor, like a spinning top, whose kinetic energy is given as

$$K = 1/2 I\omega^2, \tag{1.3}$$

where I is the moment of inertia, and

$$I = 1/2 MR^2 \tag{1.4}$$

for a uniform disk, and ω is its angular velocity. From the above, we can approximate $M = \pi R^2 H \rho = \pi \, (0.6 \text{ Mm})^2 \, (5 \text{ km})$ 1.22 kg/m^3 = 69 × 10^{14} kg, so that $I = 1242 \times 10^{24}$ kg m^2. Now, we can set $1/2 I\omega^2 = 8.18 \times 10^{18}$ J, the kinetic energy estimated above, to find the angular velocity as $\omega = 1.15 \times 10^{-4}$ s^{-1}. This rate, corresponding to a period of 2.42 h, implies that the rotational speed at radius R is $v = \omega R = 69$ m/s, comparable to the

104 m/s that was directly observed as mentioned in Fig. 1.1, as the renormalized peak speed. (How so much of the energy released from the condensation of water gets funneled into this single mode of motion is an interesting puzzle, reflecting the fact that it takes several days for the storm to get organized and up to speed in its circular motion.) We can now find this "typical" angular momentum $L = I\omega = 1.43 \times 10^{23}$ kg m/s. We can also use the power formula $P(A) = A\rho v^3/2$ derived above asking for the power represented by the wind crossing an area HR_o, where H is the height and R_o is the overall radius of the storm, assumed in solid-body rotation. So this power is

$$P = \rho H/2 \int (\omega R)^3 \, dR = \rho H/8 \, \omega^3 R_o{}^4. \qquad (1.5)$$

This evaluates as 150 TW. This is a better estimate of the wind power than the 257 TW that was obtained above because it addresses the change of wind speed with radius. We got this number by finding the rotational angular velocity and adjusting that to provide the previously found kinetic energy. The resulting wind speed vs radius rises as a straight line from the origin in Fig. 1.1 to 69 m/s at the outer radius. This is a rough approximation but not an unreasonable one. The number is about 30% of the generated power as estimated from the rainfall, that was 441 TW and we can assume that the rest of the wind power flows in the updrafts in the thunderstorms of the eyewall.

1.4   Dissipated Power from Wind into Sea

A different power related to the hurricane is the frictional or dissipative power delivered directly to the ocean, by the horizontally flowing wind. According to Emanuel (1999), this is governed by a drag coefficient C_D about 2×10^{-3} times the area integral of wind speed cubed. Thus, using $v = \omega r$, on the assumption of solid-body rotation,

$$P = 2\pi\rho \, C_D \int r dr \, v^3 = 2\pi C_D \rho \int r^4 \, dr \, \omega^3 = \rho \, 2\pi/5 \, C_D \, R_o{}^5 \, \omega^3 = 362 \text{ TW}.$$

$$(1.6)$$

This number, 362 TW, is larger than the 150 TW for the azimuthal wind power available. The difference is probably due to the increase in wind speed at the optimum altitude vs the sea surface. So, we will assume that the 104 m/s is reduced to 77.6 m/s (174 mph) at the surface, which removes this inconsistency.

Now, there remains a difference between the 441 TW estimated from the rain and the 150 TW estimated for both the achieved wind power in the circular motion and the frictional energy given to the sea, all based on the analysis of hurricane Isabel of 2003 whose outer radius we take as 600 km.

(Emanuel 1999, who quotes 100 TW from the book of Anthes for the average latent heat release, also gives his estimates between 3 TW and 30 TW for the dissipation. He takes smaller radii than we have adopted, based on Fig. 1.1 a radius of 600 km for Isabel, as well as smaller wind speed.)

The remaining difference in power, 441 TW – 150 TW = 291 TW can be attributed to vertical updraft motions in the thunderstorms that form the eyewall of the hurricane. We will see below that updraft wind power in a thunderstorm can be approximated as ≈ 2 TW, so about 145 eyewall thunderstorms would account for this extra power. The power from the rain in our picture should first go into the wind energy and then into roiling the sea, with a considerable amount sending kinetic energy into the top of the atmosphere.

The hurricane acts first as a giant humidifier and then as a rain- and wind-making machine. The average rainfall under hurricane Isabel (Fig. 1.2) is 2.5 mm/h, falling on a large area defined by the radius R, taken as 600 km. This rain would not fall in the absence of a hurricane. The hurricane persists for times approaching or exceeding a week[1], so that the cumulative rainfall under the hurricane might be 2.5 mm times 7 times 24 or 420 mm, equal to 16.5 inches.

That hypothetical 16.5 inches of rain would be distributed over the path of the hurricane, whose speed is variable.

---

[1]A great variety of growth paths are followed by real hurricanes, but all take several days, with wind speed and rainfall increasing eventually. See for example a video of Hurricane Harvey before landfall: http://cimss.ssec.wisc.edu/goes/blog/wp-content/uploads/2017/08/Harvey_band13_anim_faster.mp4.

But the notable hurricane Harvey, August 26, 2017, described in a detailed NOAA report,[2] dropped 60 inches of rain on a location near Nederland, Texas, and a nearby gauge recorded 6 inches of rain in 1 h (152 mm/h) of peak precipitation. The NOAA report shows that the wind speed and local rain rate changed together, building up at the peak of the storm, consistent with our analysis that the rain indeed is a measure of the kinetic energy release. Figure 1.3 shows a total water density image of Hurricane Harvey at landfall. This amazing image shows total precipitable water (TPW) in millimeters, in the air column above the imaged point. It was obtained from the Terra Modis satellite imaging of 11 μm infrared radiation. This shows that the airborne water density follows the general shape of the wind speed distribution, consistent with Figs. 1.1 and 1.2, but with a smaller overall radius. The actual speed distribution rises linearly from the center, reaches a peak, and then falls to zero at the outer boundary. Figure 1.3 again suggests that the hurricane is a giant humidifier, a machine to pump water vapor out of the sea, up into a ring of precipitable water in the sky around the eye. The water then falls out as rain mainly below the central ring, accompanied by fearsome winds as a release of the condensation energy. The image also shows that the eye, here about 20 miles across, at the center, is free of precipitation.

The eye, here seen as a striking feature of a mature hurricane, appears, to make the problem more interesting, to be beyond a Physics-One model. We will discuss the formation of the eye, as a feature of the hurricane phase of matter, later in this book. Actually, some smaller hurricanes do not have this feature; for example, Hurricane Marco, as of October 7, 2008,[3] reported as having a radius of 11.5 miles, or 18.5 km, was well-formed but without an eye.

So, the Physics-One model is still worth consideration, and we return to that discussion.

Next consider the effect of the rotation of the earth, whose angular velocity is

$$\Omega = 2\pi/[(24)(3600)] = 7.27 \times 10^{-5}\,\mathrm{s}^{-1} \qquad (1.7)$$

---

[2]https://www.nhc.noaa.gov/data/tcr/AL092017_Harvey.pdf.
[3]https://en.wikipedia.org/wiki/Tropical_Storm_Marco_(2008).

at the North Pole. The component of this angular velocity perpendicular to the earth's surface is reduced by a factor sin φ at North Latitude φ. As a thought experiment, imagine taking the disk-shaped mass of air that we first considered, stopping its motions, and placing it exactly above the North Pole. The air mass would sit there and see the earth rotating below it. From the point of view of the earth, correspondingly, that air mass would be rotating in the opposite direction, and its angular momentum would be $I\Omega$. This earth-rotation-induced angular momentum scales as sin φ, where φ is the North latitude, and thus disappears at the Equator.

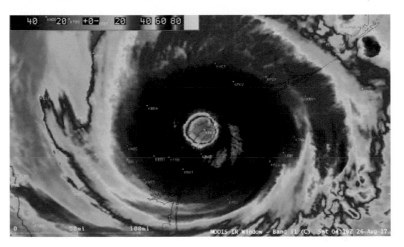

NASA TERRA MODIS INFRARED IMAGE OF HARVEY AT 0419 UTC 26 AUGUST 2017 JUST AFTER LANDFALL AS A CATEGORY 4 HURRICANE IN TEXAS. IMAGE COURTESY OF UW/CIMSS.

Figure 1.3 Infrared image, at 11 μm wavelength, shows the eye of Hurricane Harvey crossing the Texas coastline, seen as a diagonal blue line crossing the eye. The color scale is of TPW, in millimeters, the scale at the top left. Note the 100-mile scale bar at the lower left. The rain-free eye of this storm is seen to be about 25 miles (40 km) across from the length scale, while the outer radius is about 120 miles or 193 km. https://screenshots.firefox.com/ a29u9wl3lpgXNoGt/www.nhc.noaa.gov;http://cimss.ssec.wisc.edu/goes/blog/ wp-content/uploads/2017/08/170825_goes16_infrared_Hurricane_Harvey_ landfall_anim.mp4.

If we consider the Leeward Islands, the location of hurricane Isabel at its peak, at 19° North Latitude, the vertical angular velocity component of the earth rotation is 7.27×10^{-5} s^{-1} sin (19) = 2.4×10^{-5} s^{-1}. Above, we found that Isabel had an angular velocity $\omega = 1.15 \times 10^{-4}\,\text{s}^{-1}$ so its angular speed was increased by a factor of 4.8 beyond that arising simply from a rotation of the earth, and its rotational energy was increased by about a factor 23. Thus, the rotation direction and a substantial initial angular momentum are typically provided by the rotation of the earth, except at the Equator.

Hurricanes do not form near the Equator; thus it appears that the initial angular momentum is needed to form the storm. On the other hand, it seems likely that a mature storm would indeed persist if one could magically turn off the earth's rotation. If that is right, then hurricane matter will have two degenerate forms, clockwise and counter-clockwise in rotation. An aspect of the increase in angular momentum is that the radially inward airflow also carries with it angular momentum. A statement from the meteorological literature is that "the inward flow concentrates ambient vorticity," where vorticity has units of inverse time. It is also possible that the kinetic energy generation from latent heat by the storm, which is not fully understood, also acts to increase the angular momentum, accompanied by an equal and opposite transfer of angular momentum to the ocean under the storm.

It appears that above the critical temperature a runaway situation exists, at least in principle, at a location of low atmospheric pressure, in which faster inflow provides positive feedback, in one form by making the ocean surface foamy, which increases its generation of water vapor, and possibly also by making the path a spiral. That allows more path length and a better chance to raise the incoming air to saturation humidity by the time it reaches the eyewall, where the air is deflected vertically. The nature of the runaway above the critical temperature is that an increase in the inflow air speed acts to further increase that speed. The basic temperature effect is to increase the vapor pressure of the water, which, according to the Clausius–Clapeyron equation of physical chemistry increases exponentially with increasing temperature in the range involved, near 30 °C. In this range, not only does the vapor pressure increase with temperature, but its rate of change also increases. The runaway situation further reduces the atmospheric pressure in the center of the storm, increasing the inflow and vertical lofting of humid air. The atmospheric pressure is basically determined by the mass of air above the location, and that mass is reduced as the temperature of the air above is increased, reducing its density. The hurricane in full form raises the temperature of the air above its center all the way up to the top of the troposphere, leading to the extraordinarily low atmospheric pressure under the center. This reorganization of the atmosphere, from sea level to the top of the troposphere takes several days to occur and can be described as a new form of matter characterized by a warm core, low central pressure, and rapid circulation around the center.

To return to our Physics-One assessment of an exemplary hurricane, a typical tropical cyclone, we ask how this rain-making machine, spontaneously generated roughly 8180 PJ of kinetic energy, at a power level of 441 TW. To do this, we need the further fact that the atmospheric pressure at the center, under the eye, is additionally reduced, leading to a radial inflow of ocean surface air, and this further pressure reduction scales with the energy (wind speed squared) of the storm. (The initial pressure p is set by the local weather condition, and we are starting at a point of local low pressure.)

1.5 How Latent Heat Release Powers a Hurricane

The airflow conditions act to modify low pressure. The first approach to this is through Bernoulli's Principle, that is, the pressure p of an incompressible fluid reduces as its speed increases, according to the equation, for energy per unit mass:

$$1/2v^2 + gz + p/\rho = \text{const,} \qquad (1.8)$$

where ρ is the air mass density, z the height above the ocean surface, and g the acceleration of gravity, 9.81 m/s^2. This indicates that the pressure will be lower at the center, where the wind speed is higher. The inflowing air actually follows a spiral track, as suggested by spiral bands in Fig. 1.3. We can apply Eq. (1.8) along such a path leading from the outside, radius R, where v is zero and the pressure p_R is nominally 1000 mB, equal to 101 kPa, toward the center, where we can take the speed as 50 m/s and the pressure is the unknown p_0. So, we find

$$p_0 = p_R - 1/2\rho v^2 = 101 \text{ kPa} - 1.525 \text{ kPa} = 99.48 \text{ kPa} = 985 \text{ mB.} \quad (1.8a)$$

This reduction is an underestimate, since the Saffir-Simpson scale ranges from Category 1 hurricanes with a barometric pressure of greater than 980 mB (that cause minimal damage), to Category 5 hurricanes with a central pressure less than 920 mB. The pressure for the record Typhoon Tip mentioned above was 870 mB, with a maximum sea-level air speed of 87.2 m/s, corresponding to 195 mph.

But this does not fully describe the situation of the inflowing air, near sea level, toward the center of the storm. The radial inflow of air, once it reaches the center, then is directed vertically; there is an updraft at the center of the hurricane. (The flow of air is spiral, moving rapidly in the azimuthal direction, as well as vertically.)

Updrafts, in reality, centered in an annular region near the wall of the eye, as shown in Fig. 1.3, are provided in part by a ring of thunderstorms, which we will discuss later. Thunderstorms that sometimes are capable of lofting golf-ball-size hailstones have strong updrafts of buoyant warm moist air. (It is credibly reported that larger size hail, even up to soft-ball size, is sometimes seen. It is reported that in Texas, damage from hail exceeds that from tornadoes.)

While Bernoulli's law partially describes what happens, it needs to be extended to address the driving forces behind the airflows. In reality, in this radially inward, and then vertical, the spiraling flow of air, the nature of the fluid, the air, is changing, as the air picks up moisture from the ocean surface. The air, with added water vapor at the level q grams of water per kilogram of air, is a complex fluid, beyond that assumed in Bernoulli's Principle, and carries heat energies beyond the basic kinetic energy $1/2v^2$ per unit mass. Extended heat-containing energy may be called an enthalpy, and an expression used by Bister and Emanuel (1998) is

$$E = c_p T + gz + L_v q + 1/2v^2 \quad (1.9)$$

per unit mass, expressed as joules per kilogram. Here, c_p = 1004 J/kg is the heat capacity of air that arises because the molecules have internal motions of rotation and vibration, called degrees of freedom, which in thermal equilibrium come to energy values $1/2k_BT$, where k_B = 1.38 × 10^{-23} J/K. L_v = 540 cal/g, is the important latent heat of vaporization and q is the specific humidity, the mass of water, usually in grams per kilogram of air. (The specific humidity increases exponentially with temperature in the tropical temperature range, increasing more than 5% per degree. This basic property of water means that hurricane strength and damage will increase with global warming, see Irvine et al., 2019.)

These authors say that doubling CO_2 will increase the global energy dissipation, a measure of the damage cost due to hurricanes, by 17.6%. They say that most of this increase would be removed by a modest program of solar geoengineering, to slightly cool the earth.

The first three terms in Eq. (1.9) are sometimes called the "moist static energy," and this quantity increases as the air moves into the center of the hurricane along the ocean surface It is estimated, in the inflow, that the surface air may increase its specific humidity by 5 g/kg. The fact is that the moist static energy at the sea surface is much larger than in the air higher up. This disequilibrium, in energy content (enthalpy) between air at the sea surface vs. in the atmosphere above, is the basic source of energy for the hurricane. It is estimated by Emanuel (1991) that bringing the whole troposphere into thermodynamic equilibrium with the ocean would require the transfer of about 10^8 J/m² of energy from the ocean. The exemplary hurricane Isabel discussed above, with an energy of 8180 PJ and a radius of 0.60 Mm contained only 0.72% of Emanuel's estimated energy, suggesting that stronger hurricanes would be possible from the vast energy reservoir of the ocean. Another way to think of the 100 MJ/m² estimated by Emanuel is in terms of water in the atmosphere, it would correspond to about 44.3 mm of water in a vertical column.

The hurricane, like the thunderstorms that precede it, spontaneously arises to move the macroscopic tropical sea-atmosphere system toward equilibrium. (In practice, a nucleus, an initial low-pressure region is needed to start the hurricane.) The hurricane self-organizes to optimize the approach to that equilibrium, and the key step is transferring water in vapor form from the sea surface to the airflow toward the center of the hurricane.

The amount of water vapor contained within a parcel of air can vary significantly. For example, a parcel of air near saturation may contain 28 g of water per m^3 of air at 30 °C, but only 8 g of water per m^3 of air at 8 °C. So the idea is that a cubic meter of saturated air at the base of the hurricane center may rise, and lose about 20 g of water releasing 540 × 20 cal of energy when it gets to an altitude where the temperature is 8 °C. The typical rate of decrease of temperature in the atmosphere is 10 K per km, so that altitude would be around 2.2 km, where the excess water vapor may condense into rain, releasing energy. The specific humidity above the warm tropical ocean falls off substantially with altitude, and is reduced by about 25% at 1.5 km. An example of this is seen in Fig. 1.4, after Liu et al., 1991.

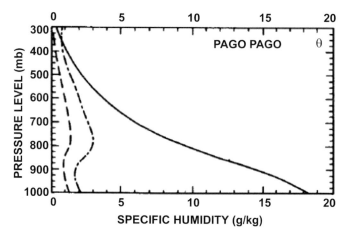

Figure 1.4 Profile of specific humidity vs air pressure (altitude) above the tropical ocean at Pago Pago, in American Samoa, at 14° S latitude. Reproduced from Liu et al., 1991. © American Meteorological Society. Used with permission.

The incoming air, starting outside and above the hurricane, is less humid, and the spiraling path toward the center of the hurricane, at levels near the sea surface, serves to raise the humidity to near the saturation level of 28 g per cubic meter (23 g/kg). It is known that the pickup of water vapor is increased by the speed of the air and also by the length of the path.

For example, the room humidifier pumps water vapor into the room by blowing air past a moist surface, with surface area increased by using a wick or filter element. The faster the fan spins the more water is put into the room air. That motion of air, not unlike that under the hurricane, promotes an equilibrium water content of the room air with water at the temperature in question.

In the hurricane, the rotational motion, increasing the inward spiral path length, and the speed of the air, both increase the rate of water, as vapor, delivered to the base of the hurricane. As mentioned, this water is thrown up into the sky and eventually condenses as rain, releasing the 540 cal/g, which feeds the rotational energy of the hurricane. It appears that added positive feedback, with increased wind speed, comes as the water surface area is increased by wave motion or even the formation of breaking whitecap waves, which add a foam of bubbles with increased surface area.

The specific humidity is the mass of water (as vapor) in the air, and this quantity, often called q, is strongly dependent on temperature, as set by the properties of water. The partial pressure of water vapor in the air is another way of describing its potential to drive a hurricane. An approximation, based on the Clausius–Clapeyron equation of physical chemistry, is

$$p = 6.109 \text{ mbar } \exp[17.67T/(243.5 + T)] \tag{1.10}$$

(with T in Celsius).[4] This pressure is 42.4 mbar, and changes by 5.9% per degree, at 30 °C. The rapidity of the increase of the vapor pressure of water with temperature is made clearer by comparing its 5.9% rise per degree with 1/303 = 0.33% rise in temperature for one degree at 303 K. The fractional increase of the vapor pressure is 17.9 times faster than that of the absolute temperature, at 30 °C.

---

[4]https://en.wikipedia.org/wiki/Vapour_pressure_of_water.

From a physics point of view, the transfer of water molecules from the liquid to the air above it is a molecular aspect. If the energy to vaporize a gram of water, with molecular weight 18 g, is 540 cal, it is about 540 cal/[(1/18) × N_A] = 540 cal/[(1/18) × 6.02 × 10^{23}] = 1.61 × 10^{-20} cal/molecule. Here, 18 is the molecular weight of water, and Avogadro's number N_A is the number of molecules in one mole. This energy is 6.76 × 10^{-20} J/molecule. In classical physics, the probability P of an energy-deficient event is often represented by

$$P = \exp\left(-E/k_B T\right), \qquad (1.11)$$

called the Maxwell-Boltzmann factor, that in this case at 303 K, would be exp(–16.167) and at 302 K would be exp(–16.22). The ratio of these two numbers is exp(0.053) = 1.054, so there is an increase of 5.4% per degree, which closely matches that mentioned above, from the Clausius–Clapeyron equation.

The rate of evaporation is crucial in powering the hurricane. One way to look at this is that the water is covered by the partial pressure of its vapor and that the wind essentially carries this vapor away with it. This is not correct for substantial evaporation, because energy is being lost to the evaporation process and, in reality, that energy has to come from the air, lowering its temperature. This effect is used in a variant of the familiar humidifier that was mentioned above, called an "evaporative cooler" or "swamp cooler." These devices, used for cooling dry room air, are common in the Southwest of the U.S., where high temperatures and low humidity are common. Such an "isenthalpic" or "adiabatic" evaporation interaction would be possible within Eq. (1.2), which contains the sum $c_p T + L_v q$, that is, suggested to be constant, based on the air cooling by the "swamp cooler." If q changes by 1 g/kg (compared to the saturation value of about 23 g/kg), the corresponding change in T would be $\Delta T = L/c_p = 540 × 4.184/1004 = 2.3$ °C. The 2.3 °C cooling is reasonable for the water cooler, but no temperature change is believed to occur in the approach of air to the center of the hurricane, this path is considered to be at a constant temperature. So, it must be that further interaction of the moistening (and thus cooling) air with the water surface keeps the air temperature close to that of the sea. So, the inflow process is considered to be isothermal and not isentropic.

The main question that we examine in this book is, how much bigger and more destructive will hurricanes become as the earth warms up? It is amazing that already observed hurricanes/cyclones have been reported with diameters ranging from 23 miles (Marco) to 1380 miles (Typhoon Tip); these are a factor of 60 different in size. The top speeds we have seen reported are 104 m/s for Isabel and 87.2 m/s for Tip. How high can these numbers go? For a start we can look at a portion of Eq. (1.8), $L_v q + 1/2v^2$, per unit mass, moist enthalpy at constant temperature and altitude, and see how large a v is available by switching from $q = 0$ to a large value. Since L is 2.259 MJ/kg, the biggest number we can find is

$$v = [q \, 2 \, L]^{1/2} \qquad (1.12)$$

that equals 2126 m/s for $q = 1$ (1000 g/kg). To get 85 m/s, we would need a change to $q = v^2/2L = 1.6 \times 10^{-3} = 1.6$ g/kg. This would represent the change from the sea surface to a layer just above, and according to Fig. 1.4, where the observed value is 18.2 g/kg at the surface near Pago Pago, a small height change for the layer just above could indeed provide a change in water content of 1.6 g/kg. The q value as we have seen is exponentially dependent on water temperature, amounting to the change of 5.9% per degree nearly up to 30 degrees.

These questions are addressed by Emanuel (1991), who suggested forms of a single scaling parameter χ to explain the range of wind speeds and storm sizes seen in hurricanes, and also to provide limits to these quantities. A form of this parameter that scales as the square of the maximum wind speed, V_m^2 is

$$\chi = (T_s - T_t)/T_s \, (E_0^* - E_a) \approx V_m^2, \qquad (1.13)$$

where subscripts s and t, respectively, indicate sea surface and top of the atmosphere. The energy E term, as shown in Eq. (1.9), the first being the value for saturated air very close to the sea surface and the second a value in the ambient above a normal ocean surface. The enthalpy energy difference term represents the disequilibrium in enthalpy, basically water content, in saturated air at the ocean surface and what is actually observed above the normal ocean. This difference may be tapped by the hurricane to reduce the disequilibrium. The temperature term derives from an interpretation of the radial motion of the air in terms of a Carnot cycle, this term represents the efficiency of the cycle with a change in the denominator temperature to the troposphere value according to Emanuel as accommodating some feedback of waste dissipation energy back into the input of the engine, increasing its efficiency. The Carnot cycle may be described as an initial approximately isothermal step, as sea surface air from outside the storm spirals into the low-pressure center, gaining enthalpy; followed by a rise to the top of the troposphere, in a constant enthalpy process (see Eq. (1.2)), where the initially large latent heat is exchanged into other forms of energy, including kinetic energy in the vertical motion, with a fall in temperature from roughly 300 K to about 200 K; followed by a high altitude expansion to a large radius in which energy is lost by radiation to space, and then a return of air to the sea surface.

It is argued by Emanuel (1991) that the maximum wind speed v scales as $\chi^{1/2}$ while the diameter of the storm scales as

$$\chi^{1/2}/f = \chi^{1/2}/2\Omega \sin \varphi. \qquad (1.14)$$

Meteorologists call

$$f = 2\Omega \sin \varphi \qquad (1.15)$$

the "Coriolis factor." It appears from this that the largest storms may arise near the Equator where the latitude is small, presumably a factor in the initial formation of the rotating storm. It seems from this that a storm starting with higher initial angular momentum, at higher latitude, may be in some way limited in its size. The temperature dependence of the whole process is made obvious by the fact that the storms only form in the warmest part of the ocean, and, there, only during the warmest part of the year. It seems likely that global warming will make the tropical storms worse, the question is how much worse. Some temperature dependence is present in Eq. (1.10) from the exponential dependence of the vapor pressure on temperature, but the question is if that is enough to explain what actually happens.

An important advance in this regard was the definition and tabulation of the power dissipation index (PDI) by K. Emanuel, 2005a.

The PDI is based on the integral over the area and duration of the storm, of the third power of its wind speed, v^3:

$$\text{PDI} = 2\pi \int\int v^3 \, r dr dt. \tag{1.16}$$

1.6 Dissipation Measurements of Emanuel

The data (Emanuel, 2007) in Fig. 1.5 show, in separate curves, for the North Atlantic, the PDI, and the sea surface temperature vs time from 1950 to 2010. The direct relation between the temperature and the PDI is not obvious from these curves. Both curves show oscillations nearly in synchronism that arise because the temperature range is very small and complicated factors that affect the detailed sea surface temperature tend to confuse the issue. The blue curve is the temperature with a large constant subtracted that allows the blue curve to include zero, matching the zero value of the PDI. The green curve is the compiled and computed PDI of historical storms.

The unit of the plotted PDI is m^3/s^2 on an annual basis, thus representing v^3 times time, nominally one year. Considering that the hurricane season is roughly a quarter of a year, 7.88×10^6 s, the value is shown on the right end of the figure, 4×10^{11} m^3/s^2, divided by that time and subjected to a cube root, which corresponds to a wind speed of 37 m/s. This is a reasonable number and represents all tropical cyclones in the North Atlantic over a one-year period near the year 2005. Again, the blue comparison curve is the sea surface temperature, in Celsius, subject to a fitting offset to line up the two curves.

The detailed fit obtained here, as well as similar fits published by Emanuel (2005a), is thus of the form:

$$PDI = \text{const.} \, (T - T_c), \tag{1.17}$$

where const. and T_c are fit parameters. The offset can be interpreted as a critical temperature for the hurricane. It is well known (Dare and McBride, 2011) that hurricanes do not commonly form unless the sea surface temperature is at least 26.5 °C. These critical temperature fit values were not disclosed in the Emanuel 2005a paper, while the 2007 paper showed 26.5 °C as the offset, but did not identify it as a critical temperature, nor recognize the employed fit equation as $PDI = \text{const.}\ (T - T_c)$.

It appears that both the PDI and the offset temperature have approximately doubled from 1990 to 2010, and it seems that the actual sea surface temperature increase over that time is about 1 degree. This temperature dependence was not emphasized by Emanuel and is not easy to visualize from Fig. 1.5.

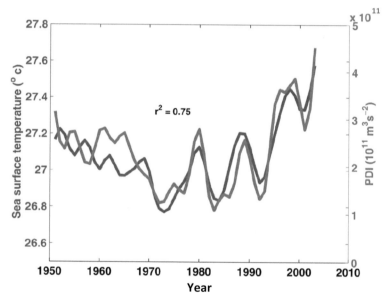

Figure 1.5 Measure of hurricane energy loss over the North Atlantic, green curve, right scale in units of 10^{11} m^3/s^2, is compiled from integral over time of the third power of velocity over historical storms. The blue curve is the North Atlantic Sea surface temperature plotted with a temperature offset. The details of these curves are amazingly superimposed. Data of Prof. K. Emanuel of MIT in 2007. (K. Emanuel, "Environmental factors affecting tropical cyclone power dissipation" *J. Climate* 20, 5497, Fig. 1a.) The underlying fit formula here, $PDI = \text{const.}\ (T - T_c)$ was not identified by Emanuel but was first recognized by Wolf, 2020.

1.7   Evidence for a Second-Order Phase Transition

The direct relation between temperature and PDI inherent in the careful data was first extracted by Wolf, 2020, by recognizing the fit algorithm (Eq. (1.17)) and simply replotting the data points from Fig. 1.5. The replotted data is shown in Fig. 1.6.

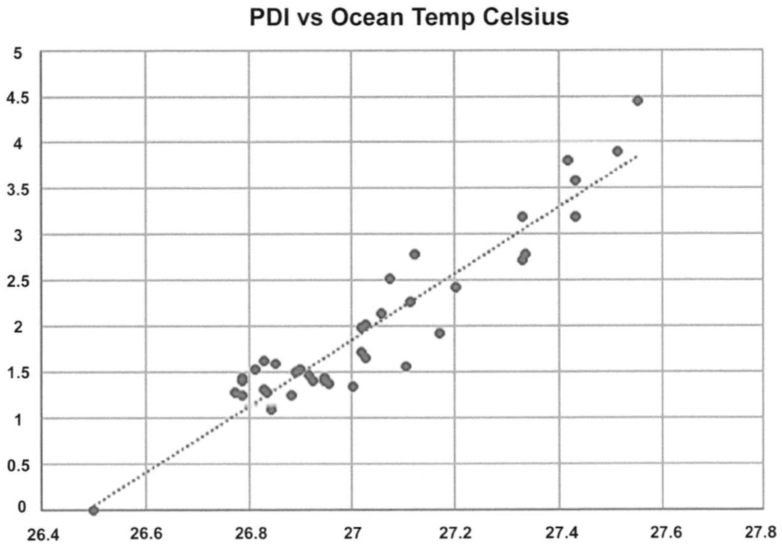

PDI vs Ocean Temp Celsius

Figure 1.6 PDI in units 10^{11} m^3 s^{-2} vs ocean surface temperature in Celsius. These are direct data points replotted from Emanuel 2007, Fig. 1a, representing the North Atlantic from 1970 to 2005. The extrapolated line matches the 26.5 °C found by Emanuel 2007, Fig. 1a, and also matches the threshold temperature found by Dare and McBride, 2011. This type of fit appears to be consistent with a phase transition of second-order as stated in the text.

1.8   Implication for Global Warming

The replotted data from Fig. 1.5, and as shown in Fig. 1.6, confirm Eq. (1.17), in the straight line shown extrapolating to the critical temperature of 26.5 °C. This line increases from PDI near 1.5 (PDI in units 10^{11} m^3 s^{-2}) at 26.85 °C to about 4 at 27.6 °C. That is a rate of change of 3.33 in PDI per degree. If we extrapolate 2 degrees higher, say by the year 2100, from the value of 4 in 2005, we find PDI = 10.66, that is, a factor 10.66/4 = 2.67 larger. In terms of wind speed, the increase will be proportional to the cube root, thus an increase in wind speed by a factor $(2.67)^{1/3} = 1.386$. So, the maximum wind scale would be increased by 39%.

In the conceptual terms of Eq. (1.17), as stated by Wolf, 2020, it is very likely that the formation of the hurricane is a second-order phase transition. The order parameter for the hurricane phase is the azimuthal wind speed *v*, which scales as the cube root of the PDI. So, the hurricane phase transition is characterized as having a critical exponent of 1/3, represented by a relation:

$$V = \text{const.} \ (T - T_c)^{1/3}, \tag{1.18}$$

which follows from the directly observed in Eq. (1.17). From the point of view of the second-order phase transition, 1/3 is a reasonable value for a critical exponent. On the other hand, the hurricane literature shows no discussion of such a temperature dependence. On the contrary, the prevalent related quantity in that literature is the "potential intensity," which has no sharp change near 26.5 °C and does not go to zero at any particular temperature. We will discuss later the possibility that the temperature dependence of the PDI is different from the potential intensity by virtue of general rules for physical quantities near the second-order phase transition.

Chapter 2

Introduction to the Tropical Atmosphere

The major part, about 80%, of the atmospheric mass, sea level up to about 12 km (Fig. 2.1) is called the troposphere. Its properties are more tied to the earth's surface, while the properties of the higher stratosphere are more set by the incoming radiation from the sun. The height of the atmosphere is on the order of 0.1% of the earth's radius, and its mass is extremely small relative to the earth.

The atmosphere is mostly molecular nitrogen and oxygen, with molar masses of 28 and 32, respectively. It can be treated as an ideal gas, that has, at sea level, a particle concentration of 2.4×10^{25} m^{-3} at the nominal temperature of 300 K and the nominal pressure, 101 kPa, also written as 1000 mB or 1010 hPa. Taking the average mass as M = 30 AMU per molecule, the mass density at ground level ρ = $2.4 \times 10^{25} \times 30 \times 1.67 \times 10^{-27}$ kg/m^3 = 1.20 kg/m^3. Some properties of the atmosphere are shown in (Fig. 2.1), after Lstiburek, J. (2014).

2.1 Pressure and Temperature vs Altitude

Considering the pressure change $\delta P = - \rho\, g\, \delta z$ with a change of height δz in the earth's gravity, g = 9.8 m/s^2, we can find the pressure as a function of height z. Thus,

Physics and Future of Hurricanes
Edward L. Wolf
Copyright © 2023 Jenny Stanford Publishing Pte. Ltd.
ISBN 978-981-4968-54-6 (Hardcover), 978-1-003-33125-4 (eBook)
www.jennystanford.com

$$\partial P / \partial z = - \rho \, g. \tag{2.1}$$

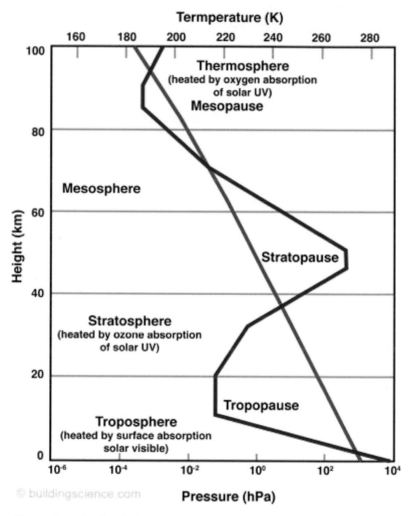

Figure 2.1 Sketch of the atmospheric temperature profile, in red. The lapse rate –*dT*/*dz* up to about 12 km is in the range of 8–10 K/km. For perfectly dry air, the rate is 9.8 K/km. Ground pressure is 1010 hPa or 101 kPa. Reproduced with permission from Lstiburek, J., 2014. © Building Science Corporation.

This is 12 Pa/m, or about 1 hPa (hectopascal) per 8.3 m. If we combine this with

$$P = \rho RT/M \qquad (2.2)$$

(the ideal gas law, where the gas constant $R = N_{Av} k_B = 8.314$ J/(mole K) and M is the molecular mass), the differential equation:

$$\partial P/P = -(gM/RT)\, \partial z. \qquad (2.3)$$

Thus, we find

$$\ln(P/P_0) = -z/h, \qquad (2.4)$$

where,

$$h = RT/gM \qquad (2.5)$$

can be called the height of the atmosphere.
   Thus,

$$P = P_0 \exp(-z/h). \qquad (2.6)$$

The height evaluates as $h = 8.48$ km for the atmosphere as a whole, but, e.g., is reduced to 5.78 km for CO_2, which has a mass of 44 compared to the atmospheric mass of 30. So at the height of the airliner, 30,000 ft or 10,000 m, the air pressure will be 101 kPa exp($-10,000/8480$) = 31 kPa.

Compared to the radius of the earth, 6371 km, the atmosphere's height of 8.5 km is 0.13%, small indeed. The mass of the atmosphere is 5×10^{21} kg, which is an even smaller fraction, 0.084%, of the mass of the earth, i.e., 5.97×10^{24} kg. The derivation we have made assumed a constant temperature, which is incorrect but not badly enough to invalidate the pressure prediction. In fact, the temperature falls with heights up to about 12 km. If we guess that the temperature T varies linearly with the pressure P, i.e., as $T = 300 \exp(-z/h)$, so at 10,000 m, the temperature would be reduced by the same factor as the pressure, we get $T = 92$ K. This guess is seriously wrong, the temperature at 10,000 m is approximately 220 K. A correct approximate understanding of the "lapse rate" $-dT/dz$, the temperature variation with height z, can be obtained from the constancy of the quantity $c_pT + gz$, called the "dry static energy" (Randall, 2012), where the specific heat at constant pressure $c_p = 1004$ J/K kg.

$$\text{Thus, } T(z) \approx 300 - gz/c_p, \tag{2.7}$$

which we can expect to be nearly correct for dry air up to about 10 km, in the troposphere, which holds about 80% of the atmospheric mass. (The temperature, see Fig. 2.1, reaches a minimum value, 218 K, is nearly constant between 12 km and 20 km, and rises at an altitude above 20 km, due to absorption of sunlight by ozone, O_3. The temperature at the top of the stratosphere, ≈ 50 km, is around 270 K.)

Equation (2.7) gives for altitude $z = 10$ km, as an estimate

$$T(10 \text{ km}) = 300 - 9.8 \ 10,000/1004 = 202.4 \text{ K.}$$

This is not too far off, and the discrepancy with the stated 220 K (Randall, 2012) exposes the fact that $c_pT + gz$ is only approximately constant, and actually increases with height.

The ideal gas law continues to apply locally, so we can use our approximate results to get an air density, proportional to $1/V = P/RT$, for the assumed dry conditions:

$$\rho(z) \approx 1.2 \text{ kg/m}^3 \exp(-z/h)/[1 - gz/300c_p]. \tag{2.8}$$

This analysis applies to air with no water vapor. Allowing humidity and the chance for cloud formation makes the situation more complicated.

2.2 Convection Generates Air Mass Motion and Kinetic Energy

The common observation is of a hawk or eagle circling effortlessly above a warm field. The bird seeks locations where the air is rising because it is warmer and therefore less dense than air in the local environment. Imagine a parcel of air, volume V, with reduced density ρ' and weight $gV\rho'$. By Archimedes Principle, it will be buoyed upward by a net force

$$F = gV(\rho - \rho'), \tag{2.9}$$

and will therefore be accelerated upward by acceleration

$$a = F/\text{mass} = F/\rho'V = g(\rho - \rho')/\rho' = g\,(T' - T)/T. \tag{2.10}$$

Here T' and T, respectively, are the Kelvin temperatures inside and outside the parcel of volume V. The last step in Eq. (2.10) comes from the ideal gas law, where density is inversely proportional to temperature. Suppose the temperature difference between the air above the ploughed field and above the woods adjacent is 3.5 °C, then the upward acceleration d^2z/dt^2 of the block of air whose area might be $L^2 = (1000 \text{ m})^2$, is plausibly

$$a = 9.81\,(3.5/300) = 0.11 \text{ m/s}^2 \tag{2.11}$$

The vertical velocity becomes $v = dz/dt = at$, with zero velocity to start, so the parcel height increases as $z = 1/2\,at^2$. Combining these last two equations gives $v(z) = (2az)^{1/2}$.

At $z = 200$ m, plausibly the height of the circling bird, i.e., 5.5 m/s, is likely enough to keep birds gliding at its height. This is a simple example of a "buoyant vertical plume" of air rising by convection. If the temperature difference was to persist to $z = 10$ km, a situation at the top of the troposphere is called "deep convection", the final air speed, neglecting any kind of friction, would be 46.9 m/s, a large speed. But this is not impossible, considering that Hurricane Isabel was reported with a direct measurement of wind speed of 104 m/s. We know that hailstones, the size of golf balls and larger, must be lifted by such updrafts in thunderstorms to provide their observed ring structure, layers like an onion. It appears that all the high speeds in storms are driven by convection. Above the ploughed field, surrounded by forest, a bird can spend many minutes, counting on the steady flow of warm air upward. So, there must be counterflow, a downward flow above the surrounding forest that then moves horizontally onto the ploughed field. So, the temperature gradient between the ploughed field and forest is pumping warm air vertically, and also creating a wind pattern at the ground level.

We calculate that the air rises according to the law $v = (2az)^{1/2}$ neglecting any kind of friction. A drag force will actually occur proportional to the area L^2, and increase as the square of the wind speed v. So, there will be a terminal velocity when the drag force equals the buoyancy force, i.e., proportional to L^3. So, the relative importance of the drag force falls to $1/L$. In atmospheric updrafts, the size scale L is so large that the air motion is not really affected by such drag forces, whose relative importance scale as $1/L$ to become vanishingly small.

In the other limit of strong friction, objects like golf ball size hail and tiny water droplets forming a cloud layer the drag force is more important limiting the speed and even holding tiny objects stationary, a class of tiny particles called aerosols.

2.2.1   An Ideal Convective Updraft

The parcel of dry air over the ploughed field on a sunny day is not likely to keep rising, it takes water vapor in the parcel to do that to achieve "deep convection". In the dry air case, as the parcel rises, its internal temperature falls to match the fall in the outside air temperature, assuming adiabatic conditions, i.e., no transfer of heat into the parcel, which might otherwise be provided by condensing moist air to water droplets, forming a cloud. Thus, the parcel expands and its invisible flexible boundaries do work on the exterior environment, lowering the parcel's internal temperature. Once the parcel temperature matches the environmental temperature the upward acceleration stops. In reality, then the temperature changes with the height z of the parcel, and of its environment become very important.

The actual T and T' temperature profiles can be incorporated into the analysis, by recasting Eqs. (2.10 and 2.11) as:

$$v^2(z) = (2az) = 2\,g(T' - T)/Tz \qquad (2.12)$$

and taking the differential

$$d[v^2(z)] = (2a \, dz) = 2 \, g(T' - T)/T \, dz, \qquad (2.13)$$

leading to the result

$$v^2(z) = (\int 2a dz) = 2 \, g \int (T' - T)/T \, dz = 2 \, CAPE, \qquad (2.14)$$

which gives the parcel speed squared, after rising through the temperature profile to altitude z, and neglecting any kind of friction.

Half this quantity is called the convectively available potential energy (CAPE) in joules per kilogram.

This is the essential concept and process for converting temperature increases from the latent heat of other sources into macroscopic kinetic energy.

Measurement of the "lapse rate" is a basic procedure in meteorology, using dropsonde sensor devices. Such devices are usually carried aloft by a weather balloon or they may be dropped from an airplane. For dry air, as we have considered, the dry adiabatic lapse rate is g/cp = 9.8 °C/km. This would apply to the dry air warmed above the imagined sunlit ploughed field in our example. Many different behaviors will occur if the parcel of warm air contains water vapor.

Observations generally give smaller lapse rates for the troposphere. In addition, at the top of the troposphere, about 11 km, the lapse rate sharply falls to perhaps 5 °C/km, stopping further vertical acceleration. However, in the noted absence of friction, the speed of the upward parcel is not changed when the acceleration goes to zero. The top height of an updraft of an area of 1 km^2 involves a moving mass on the order of 1 billion kg, not an easy thing to stop even after the acceleration is reduced to zero. This relates to the observation of "overshooting tops" of large clouds, as mentioned below.

2.3 Aspects of Cloud Formation in a Humid Atmosphere

The lapse rate, $-dT/dz$, in meteorological terms, from Eq. (2.7) is g/c_p, which is 9.8 K/km under dry adiabatic conditions. The observed lapse rate is variable, often reduced to about half this value near the earth's surface in locations where the atmosphere is moist, and this has to do with warming, as moist air rises and condenses to form a cloud. In addition, since the molecular mass of H_2O is only 18, vs 28 and 32 of oxygen and nitrogen, respectively, a parcel of moist air is slightly less dense and will be buoyed upward by surrounding dry air. The "specific humidity" is defined as the mass fraction of the air due to water vapor, ρ_{vapor}/ρ, i.e., typically parts per thousand, g/kg. The vapor pressure of water in the tropical temperature range is an increasing exponential function of temperature, which is purely a molecular property. In a crude approximation valid only near 288 K, the function is roughly

$$P_{sat} \sim 1.71 \text{ kPa } \exp[(T - 288)/16.5]. \qquad (2.15)$$

The pressure of water vapor in equilibrium at the accepted average earth surface temperature of 288 K is about 1.7 kPa, compared to the 101 kPa of the atmosphere, but this 1.7 kPa increases by 6% per K at 288 K. Above a warm tropical ocean, the saturation vapor pressure may approach 4 kPa, near 28 °C, which would correspond to about 2.5% by mass, 25 g/kg, of the atmosphere as water vapor. The relative humidity is defined as the water vapor pressure compared to the saturation value, 100% relative humidity, which implies the formation of tiny droplets, too small to fall downward, thus creating a floating cloud. The bottom of the cloud is where the water vapor in the rising parcel reaches its dew point.

To form a cloud, warmed less dense moist air continues to rise in an environment of cooler more dense air. The temperature falls with increasing altitude z, and the bottom of the cloud occurs where the available water vapor pressure reaches the saturation vapor pressure, which will fall with increasing altitude. So, the cloud bottom is higher when the humidity is lower.

2.3.1 Stokes' Law and Levitation of Aerosol Particles

The cloud contains tiny water particles, but they are too small to fall and remain suspended. The longtime suspension of small particles in the air is a consequence of the viscosity of air, in the regime of motion governed by Stokes' Law.

Namely, the force needed to move a sphere of radius R at a velocity v through a viscous medium is given by Stokes' Law:

$$F = 6\pi\eta Rv. \tag{2.16}$$

This is valid only for very small particles and small velocities, under conditions of streamlined flow. The relevant property of the medium, air, is the viscosity η, defined in terms of the force $F = \eta vA/z$ necessary to move a flat surface of area A parallel to an extended stationary surface at a spacing z and relative velocity v in the medium. The unit of viscosity η is the Pascal-second (one Pascal is a pressure of 1 N/m^2). The viscosity of air is about 0.018×10^{-3} Pa-s, while the value for water is about 1.8×10^{-3} Pa-s. The traditional unit of viscosity, the Poise, is 0.1 Pa-s in magnitude.

The fall, under the acceleration of gravity g, of a tiny particle of mass m in this regime is described, following Stokes' Law, by a limiting velocity obtained by setting F mg This gives, for small particles in the air:

$$V = mg/6\pi\eta R = 2/9\ R^2\rho\,g/\eta = 1.21 \times 10^5\ R^2\rho. \tag{2.17}$$

For example, a particle of 10 μm radius and density of 2000 kg/m^3 falls in the air at about 24 mm/s, while a 15 nm particle of density 500 kg/m^3 will fall in the air at about 13 nm/s. The latter example resembles a tiny soot particle known to come from jet engine exhaust and nucleate water vapor in the familiar contrail from the jet airplane. On this basis, a raindrop of radius $R = 0.2$ mm will fall at 5 m/s, i.e., is about right for a large raindrop.

The formation of the cloud depends upon nucleating sites for water molecules to populate and turn into larger droplets and eventually into raindrops. In a pure atmosphere with only nitrogen and oxygen molecules plus a few water molecules, the chance of two water molecules colliding and sticking together is very small, which is negligible. In practice, a water molecule finds a nucleating surface on a cloud nucleating particle, where the water molecule finds some binding energy. The water molecule will be able to hop around on that surface. In that situation, a second water molecule hopping on the nucleating surface may find the first one and the pair of water molecules on the nucleating surface can grow to lead to a liquid water drop or a snowflake if it is cold enough. The rate of forming a new cloud when the local humidity exceeds the saturation humidity depends on having plenty of nucleating sites available. The nucleating particles are typically aerosol particles that can be natural in origin, as tiny dust particles from the Sahara Desert or salt particles coming from saltwater spray, or man-made as coming from sulfur particles starting from SO_2 that is present from sulfur impurity in coal burned in electric power plants.

Returning to the warm parcel considered above as floating a hawk above a ploughed field, if the parcel has humid air, say specific humidity q = 10 g/kg, as might be encountered above Pago Pago as illustrated in Fig. 1.4 above. The acceleration given by Eq. (2.11) will repeat until the parcel cools to the dew point of the enclosed gas. At that point, the base of a cloud, heat will be released according to the formula:

$$dQ = L_v \, dq = c_p \, dT. \tag{2.18}$$

The result is that the parcel will be warmer as it rises, reducing its lapse rate. It may well still rise steadily but at the reduced lapse rate, now called the "moist adiabatic lapse rate" reflecting continued condensation as the temperature is lowered with parcel rise, but now less rapidly. The internal processes will keep the rising parcel at its dew point until the humidity goes to zero, when the air in the parcel is dry, all of its water having been turned to liquid phase and fallen away. All of this is described by Eq. (2.14) and the result is often that the parcel will rise to the very top of the troposphere in the internal environment of the thunderstorm.

Chapter 3

Deep Convection in Thunderstorms

Cumulonimbus (derived from the Latin word "cumulous" which means heaped and "nimbus" means rainstorm) clouds are tall electrically charged clouds also known as single-cell thunderstorms. We will take the horizontal scale of these clouds as a 3 km radius, larger than the ploughed field considered earlier, but much smaller than a hurricane. Such a storm acts as an upward pump of air driven by convection. If we consider the kinetic energy of a 1 km thick disk of air, a radius of 3 km moving at a speed of 46.9 m/s (see Eq. (2.12)), we find $K = 1/2mv^2 = 3.8 \times 10^{10}$ J. The corresponding power flows upward, from Eq. (1.2), $P = A\rho v^3/2$ is 1.78 TW. Although our assumed upward speed is somewhat arbitrary, these clouds are powerful engines to raise warm air and energy upward from the surface of the earth. The energy is initially mostly in the form of the latent heat of the water. The "cumulus instability" (Fig. 3.1) is described as the tendency of humid air to float upward under the influence of positive buoyancy generated through parcel temperature T' increase, from the release of latent heat. Following Eq. (2.14), we find:

$$v(z) = (\int 2adz)^{1/2} = [\, 2\, g \int (T' - T)/T\, dz]^{1/2} = (2\, CAPE)^{1/2}, \qquad (3.1)$$

which shows a vertical updraft velocity within the cloud if the temperature difference $T' - T$ continues to a high altitude.

Physics and Future of Hurricanes
Edward L. Wolf
Copyright © 2023 Jenny Stanford Publishing Pte. Ltd.
ISBN 978-981-4968-54-6 (Hardcover), 978-1-003-33125-4 (eBook)
www.jennystanford.com

3.1 Cumulonimbus Thunderstorm

The rising warm humid air forms a cloud towering as high as 12 km, as suggested in the center and right panels of Fig. 3.1. Following Fig. 2.1, the lapse rate $-dT/dz$ approaching 8 K/km extends up to 12 km, where it falls sharply to a lower value, and at this point, the rising air lost its humidity and stops the condensation, and the cloud tends to move horizontally creating what is called the anvil, a flat top of the cumulonimbus cloud. The strong central updraft of the single-cell thunderstorm is accompanied at lower levels by precipitation, such as heavy rain and lightning, producing thunder that occurs only when ice crystals are present. Heavy rain occurs as droplets in the initial condensation collide with each other to form larger drops that increasingly rapidly fall downward, as suggested by Eq. (2.17),

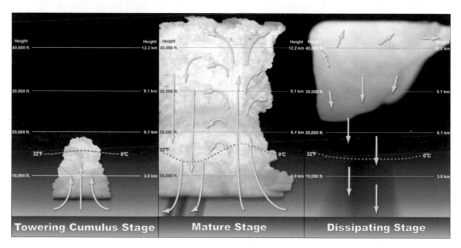

Figure 3.1 Three stages of the single-cell thunderstorm, the cumulonimbus cloud. The left panel shows warm and moist air rising by Archimedes Principle. The middle panel assumes continued temperature rise in the central updraft, also showing rain and downdrafts. The final stage occurs on a time scale of less than an hour, the cold downdrafts remove the original warm ground-level temperature, so the event is terminated. The process has sent energy vertically into the stratosphere as updraft speeds near 50 m/s at the top of the cloud overshoot upward and some moisture is carried away at a high altitude in the anvil cloud. Tstorm-tcu-stage.jpgTstorm-mature-stage.jpgTstorm-dissipating-stage.jpg, Public Domain, https://commons.wikimedia.org/w/index.php?curid=10570783

creating, by their drag effect, a downdraft. The downdraft may accelerate downward as large drops, entering warmer air, partially evaporate, leading to cooling by evaporation and downward acceleration also following Eq. (3.1), capable of giving negative acceleration values.

3.2 Proofs of Strong Vertical Air Motions in Tall Clouds

We estimated in Eq. (2.10) that a convective parcel with a continuing temperature difference $T' - T$ of 3.5 °C will reach a vertical speed of 46.9 m/s by 10 km. How can we be sure of such an idea? Direct measurement of such motion is plotted in Fig. 3.2, Bluestein et al., 1988.

The measurements were made on May 7, 1986, at three locations near Canadian, Texas, including an automobile unit that drove under the center of the storm cloud to release a radiosonde carried by a helium-filled weather balloon. Such a measurement is called a sounding, and at the same time soundings were made at nearby fixed stations in Wheeler, Texas to the south and Taloga to the east. The latter two soundings were used to establish the temperature profile and lapse rate of the "environment" of the parcel representing the major updraft in the tornado-producing storm cloud. It was found that the temperature difference $T' - T$ was about 10 °C, and that the warm parcel in the thundercloud accelerated to at least 60 m/s as it rose. The neighboring locations Wheeler and Taloga were hot and dry and the lapse rate at each could be characterized by the dry adiabatic rate corresponding to 104 F at the ground level.

A helium-filled weather balloon released from the ground directly under a huge black cloud was tracked in its vertical motion, carried along by the updraft wind, and all the time making temperature measurements. The data in the solid line reach 49 m/s by altitude 7.2 km above mean sea level (MSL). This is a direct measurement. A correction is made for the ascent speed (terminal velocity) of the balloon in still air, i.e., 4.5 m/s, which is subtracted from the center of mass velocity deduced from the radio signals giving the height. The measured quantities included temperature, pressure, and humidity; the latter was made with a "carbon hygristor". The total time for the ascent of the balloon was estimated as 440 s to reach 7.2 km. The corresponding dashed line is based on the formula $v = (2\ CAPE)^{1/2}$.

A comparison of the slopes of the dotted (adiabatic) and solid (moist adiabatic) curves suggests the moist adiabatic rate, which is on the order of half the adiabatic rate, is known to be 9.8 K/km. In the right panel of Fig. 3.3, the arrows show that the temperature difference between the cloud parcels, on the right and the environment on the left is near 10 °C for a range of altitudes above that of 500 mB. This allows an estimate of the upward convective acceleration

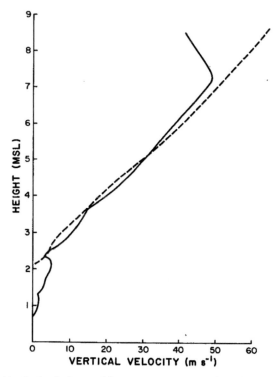

Figure 3.2 Vertical wind speed in meters per second is shown to rise with altitude in the direct measurement under and inside a thunder cloud. The solid line is measured from radiosonde signals while the dashed line shows the speed profile based on parcel theory (Eq. (3.1)) given the measured temperature difference between the parcel and the environment. Reproduced from Bluestein et al., 1988. © American Meteorological Society. Used with permission.

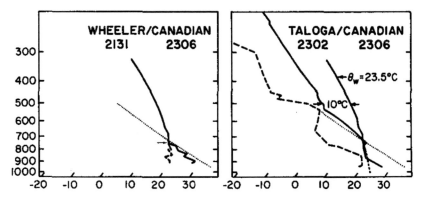

Figure 3.3 Altitude in mB (log scale) vs temperatures (solid lines) in Celsius. Composite soundings Wheeler-Canadian and Taloga-Canadian establish the temperatures and lapse rates of the environment as compared to the parcel inside the thunderstorm. Left panel: solid line lower right is from Wheeler, Texas, which matches the dry adiabatic line for 313 K (104 F), dotted line, up to 736 mB, where data cease. The arrow at 736 mB indicates the bottom of the thunder cloud. A solid line above 736 mB is measured inside a Canadian, TX, thunderstorm. Right panel: left solid line is the temperature measured above Taloga, which reasonably matches the dry adiabatic line (dotted) up to 500 mB. The right solid line is the temperature measured inside the cloud, as in the left panel, sharply warmer (10 "C) than the environment by 500 mB altitude. The cloud temperature curve is extended to intercept ground level at 23.5 °C in "potential temperature", a technical term defined as the temperature of the parcel would have adiabatically transferred to ground pressure. The potential temperature is a conserved quantity for adiabatic processes and applies to the whole inside of the cloud temperature profile. A dashed line is a dewpoint, at Taloga, the environmental point is much cooler than the observed temperature, confirming a dry and warm environment around the Canadian, TX, storm. Reproduced from Bluestein et al., 1988. © American Meteorological Society. Used with permission.

$$a = g\,(T' - T)/T \qquad\qquad (3.2)$$

inside the cloud as 0.313 m/s^2. Following the constant acceleration formula for vertical speed, $v = (2az)^{1/2}$, taking $v = 0$ at $z = 2$ km (see Fig. 3.2), one predicts $v = 57.0$ m/s at $z = 7.2$ km. This seems reasonable as the right panel in Fig. 3.3 shows the temperature difference as slightly less than 10 °C at some altitudes.

As in the earlier case above the ploughed field, the vertical velocity will continue to higher altitudes because the only parameter that goes to zero is the acceleration. So, the cloud has a vertical fountain of air and kinetic energy spewing out at its top. We have not seen evidence of friction in dealing with the large flows of air.

The dark and threatening storm cloud above the release point, where the researchers' car was parked, was probably a super-cell thunderstorm, rather than a single-cell thunderstorm, as mentioned in Fig. 3.1, but the path of the balloon was clearly confined to a single updraft. The overall storm spawned 5 tornadoes; the release of the balloon relevant to Fig. 3.2 was just after the fourth tornado had dissipated. A nearby storm also released "soft-ball size hail" stated as 7 cm in diameter at the same time, and the storm shown in Fig. 3.2 released baseball size hail, although that likely did not occur where the radiosonde measurements were performed. We take the hail as 7 cm, stated as the diameter of a baseball. According to Heymsfield et al., 2018, an estimate of the terminal velocity for a hailstone of diameter D (in cm) is

$$V = 6.35\, D^{0.97}\, \text{m/s} \tag{3.3}$$

that gives 41.8 m/s for D = 7 cm that is about baseball diameter. This is of the same order as the directly measured updraft speed of 49 m/s at 7.2 km height above sea level. Hailstone formation really requires an updraft speed larger than the terminal velocity because its layered form indicates multiple transits up to a high altitude where it is cold enough to form ice. The hail size suggests the updraft may have been stronger, the updraft measurement was unreliable above 7 km presumably because of ice formation on the airborne sensor, and the cloud likely was taller than 7 km. Updraft speed reaching 60 m/s is shown in the dashed line, based on the CAPE formula, for which the needed parameters were obtained by the sensors on the radiosonde. The nature of the "wall cloud" into which the radiosonde balloon was released is suggested by Fig. 3.4, also from Bluestein et al., 1988. This figure shows the fourth tornado produced by the same cloud as studied in Fig. 3.2, at a time less than an hour later than the picture. The tornado indicates that rotation about the vertical was present in the "tornadic" cloud, and that the angular momentum became concentrated episodically (there were five tornadoes in sequence) at the bottom of the cloud to form the tornadoes.

Figure 3.4 Telephoto picture of the third tornado formed under the Canadian, Texas, storm relating to Figs. 3.2 and 3.3. The tornado is described as a pendant from the wall cloud of the storm, seen at the top, and the radiosonde was released later under that same wall cloud, from which baseball size hail was also observed. Bluestein et al. found that the temperature inside the wall cloud, extending many kilometers upward, is about 10 °C warmer than the "environment" because of the release of latent heat from humid air in the cloud (Photograph by Howard B. Bluestein). Reproduced from Bluestein et al., 1988. © American Meteorological Society. Used with permission.

Some ideas as to how angular momentum about the vertical can be generated in such thunderclouds are shown in Fig. 3.5. The red circling line in the mid-right of the panel shows the updraft mingling with wind flowing in from the west, thereby taking on some rotation about a vertical axis. It also shows air flowing vertically out the top of the storm presumably at the high speed above 40 m/s that comes from the convective acceleration formula shown in Eq. (3.2) and also shown in Fig. 3.2.

The storm imaginatively sketched in Fig. 3.5 is shown as moving to the right, with a gust front and forward flank downdraft in front and a rear flank downdraft toward the rear. The storm is shown as extremely tall, so most of the structure would be below zero centigrade. The hailstones mentioned in connection with the Canadian, Texas storm of Figs. 3.2–3.4 of course imply that much of the storm was below freezing. While it is mentioned that the central updraft location, where the weather balloon was released may be free of rain, the large hailstone can be formed only if it accumulates mass as it goes upward in the central updraft, although it may be thrown outward and come down in downdrafts either forward or rear of the main updraft. The layered structure of hailstones clearly shows that multiple transits up and down by kilometer distances must be possible in such large storms.

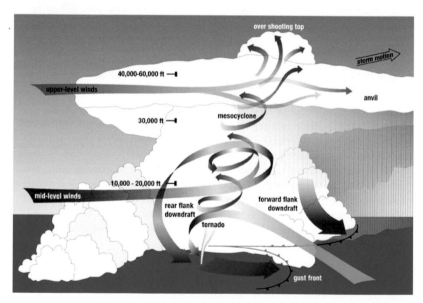

Figure 3.5 Sketch of a tornadic thunderstorm. Central updraft is shown in red, acquiring rotation by addition of horizontally moving airmass (middle-level winds, lower left) drawn into vertical motion and later into the rear flank downdraft. "Overshooting top" indicates upward release of vertically convected air as described in the text. The region under the main updraft is dry and most of the rain comes from the anvil topping cloud on the right side. https://commons.wikimedia.org/wiki/File:Tornadic_supercell.jpg

3.3   Implications of Large Hail

A 53.5 mm diameter hailstone was produced by a super-cell thunderstorm in northwest Calgary Alberta on June 13, 2020. (https://www.washingtonpost.com/weather/2020/06/15/calgary-hail-storm/). The storm produced several inches of smaller hail at nearby locations, as well as larger hail like that reported. Hail is large enough to break automobile windshields and strip siding from residences. The terminal velocity for this size, see Eq. (3.3), is 32.3 m/s or 72.3 mph, a very damaging prospect.

Again, looking at the energetics of the thunderstorm, if we take the updraft speed as 46.9 m/s and the radius of the storm as 3 km, we find, for a disk 1 km thick, $K = 1/2mv^2 = 3.8 \times 10^{10}$ J. The corresponding power flow upward, from Eq. (1.2), $P = A\rho v^3/2$ is 1.78 TW. These clouds are powerful engines to raise warm air and energy upward from the surface of the earth. The convection process allows microscopic energy in the form of latent heat to be transformed into kinetic energy of large masses of air.

3.4   The Derecho, a Moving Wall of Thunderstorms

A basic unit in meteorology is the single convective cell. The cumulonimbus thunderstorm, as shown in Fig. 3.1, arises as warm moist air rises into a cooler region above and loses its water content by condensation and release of precipitation and heat. As suggested in Fig. 3.1, the result is a correction, the air at the bottom is no longer so moist and a stabler quiescent state has been reached. The thunderstorm is a single self-correcting event that we have suggested has a lifetime on the order of an hour.

Long-lasting convective cells would require restoration of the excess humidity at the base and of the cool dryness at the top of the cell. Two situations allow long-lasting convective structures. One is the hurricane, where a group of thunderstorms arrange in a circle over water and draw in moist air from the surrounding region.

The second is the Derecho ("straight", in Spanish), where a wall or line of thunderstorms, containing wind gusts at least 75 mph, moves directly across a region, progressively removing the ground humidity and dryness aloft as it moves along, leaving a path of rainfall and wind damage.

Figure 3.6 below shows twelve rainfall images on an hourly basis of a Derecho crossing from west to east across Iowa, Illinois, and Indiana on August 10, 2020. Iowa, the central state in the picture, is 200 miles from north to south, so the Derecho wall of storms, shown in the picture by the twelve rainfall images is roughly 100 miles long, north to south, lengthening to 200 miles as it moves east. It takes 5 h to cross Iowa, whose width is 310 miles from the Missouri River to the Mississippi River, so its speed moving eastward is about 62 mph.

August 10, 2020 Derecho: Lowest Angle NWS Radar Reflectivity at One-Hour Time Steps
All times in CDT

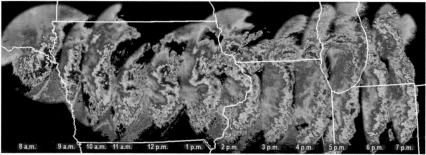

This long-lasting, severe wind thunderstorm complex (known as a derecho) produced hundreds of reports of wind damage along with numerous tornadoes.

NWS Chicago | weather.gov   Aug 11, 2020

Figure 3.6 Hourly progress of linear Derecho storm complex across Iowa and Illinois on August 10, 2020. The radar reflectivity image shows rainfall. At 8 AM on the left, the storm is in Nebraska crossing the Missouri River into Iowa. It moves eastward steadily for 11 h, becoming longer from north to south to about 200 miles. The winds in this storm were estimated at up to 140 mph, similar to a category 3 or 4 hurricane (Capucci, 2020b). https://www.washingtonpost.com/weather/2020/08/19/iowa-derecho-hurricane/.

The Derecho is defined by winds exceeding 75 mph, and also has its propagation speed, a separate quantity. The relation between these two numbers is not entirely clear, but it would appear that the faster, the complex of thunderstorms moves the greater power it might have, since the power is clearly related to the latent heat release that can be achieved per unit of time.

In the Derecho illustrated, the propagation speed is around 60 mph from the image, and the reported maximum wind estimate was up to 140 mph. Derecho storm complexes are difficult to predict and are therefore more dangerous. The necessary ingredient beyond the basic instability of the air (the excess in humidity and temperature at the ground vs their values at a high altitude), is not well known. Rare additional conditions are needed that allow such a large linear complex to form and propagate.

One set of relevant data that is known is shown in Fig. 3.7, a map of the frequency of high-speed wind incidents across the United States. The events here are simple, a sufficiently long measurement of wind speed exceeding 74 mph near the ground level.

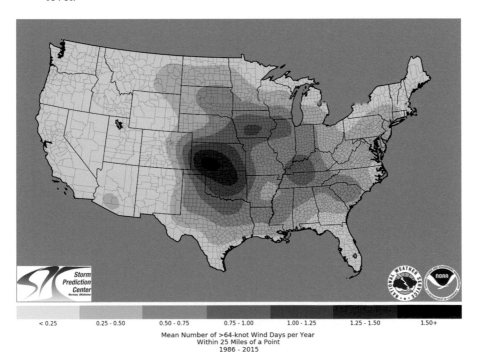

Figure 3.7 Annual frequency of wind observations faster than 74 mph in wind/thunderstorms (Storm Prediction Center of the National Weather Service). 74 mph is the formal threshold for hurricane wind. "Iowa is as prone to destructive derechos as Florida is to hurricanes" *Washington Post*, Aug. 19, 2020. https://www.washingtonpost.com/weather/2020/08/19/iowa-derecho-hurricane/.

The map in Fig. 3.7 might be called a "map of tornado alley" since tornadoes are offshoots of thunderstorms with high winds. This map is purely data, there is no theory involved. The broad, but well-defined, peak of the distribution is in southwestern Kansas, which lies south and west of Iowa, where the cited Derecho was primarily located. One can speculate that this location maximizes the basic driving instability leading to a thunderstorm, a tornado, and a hurricane: the difference, the disequilibrium between the temperature and humidity at the ground level vs their values in the atmosphere above. The ground level temperature times humidity factor alone might put the peak closer to the Gulf of Mexico and the cool dry upper atmosphere factor might put the peak nearer to Colorado, on the west side of Kansas, where cool dry higher altitude winds flow in from the Rocky Mountains. But the result of the two main factors represents a compromise perhaps midway between Louisiana and Colorado, and provides a striking image of tornado alley. One would think that this broad but well-defined distribution might be reproduced by the theoretical "potential intensity", the maximum possible hurricane speed that first principles will allow for a hurricane. That theory, to be discussed in the next chapter, might well describe the actual data here, even though the use of the "potential intensity" theory has been largely limited to ocean locations, and its predictions for the continental US are not known to the author.

The initial condition for the Derecho is an area, a region, to be swept out, of atmospheric instability, which will be in some part removed as the storm line moves through. In the case of the hurricane, the assumption is the warm ocean below represents an inexhaustible source of humidity that is controlled only by the surface temperature of the ocean and is not strongly diminished by the appearance of the hurricane.

Chapter 4

A Hurricane as a Ring of Thunderstorms

To be clear, the idealized hurricane, closely approximated by Hurricane Harvey in Fig. 1.3, and argued in this book to be a separate phase of matter, is more than a ring of thunderstorms: there is no indication in that detailed picture of separate regions. But a hurricane in a formative stage often does show separate centers, a thought process for the formation of a hurricane that can start from such an idea. The thunderstorm forming over the tropical ocean is an initial step toward reducing the lack of equilibrium between the sea level humidity and enthalpy and their values in the upper atmosphere, which was quoted earlier as representing an energy cost of 100 MJ/m². At sea, the thunderstorm shown in Fig. 3.5 is a tall tower with an updraft spouting from its top, the air will find its way back down and be recirculated upward. So, we can think of a downdraft surrounding the thunderstorm. While the air may recirculate a few times inside a thunderstorm, allowing the formation of hailstones, the thunderstorm does not perpetuate over land or the ocean for more than an hour or so. The downdraft and heavy rain at the end of the thunderstorm discourage its immediate reformation. A hurricane does not form over land; however, it is a long-lived organization of matter. The evidence is that when the ocean is warm enough, creating the hurricane structure, roughly starts by aggregating thunderstorms, and this provides a further driving force to lower the overall energy.

Physics and Future of Hurricanes
Edward L. Wolf
Copyright © 2023 Jenny Stanford Publishing Pte. Ltd.
ISBN 978-981-4968-54-6 (Hardcover), 978-1-003-33125-4 (eBook)
www.jennystanford.com

4.1 A Ring of Thunderstorms

So, conceptually, thunderstorms are arranged in a ring of variable radius. If the ring radius is very large, there will be scarcely any effect; each storm will simply have two neighbors and will pull in air from the surrounding sea surface to pump upwards. But when the ring radius is small, there will be a shortage of inflow air from the circular area inside the ring. This will occur when the returning air from the top of each storm no longer mostly falls inside the ring. The thunderstorms work as pumps for moving air vertically, pulling in from the bottom and sending it out to the top. If the ring of updraft-producing thunderstorms is small enough, there will be low pressure on the inside of the ring, and this organization of the thunderstorms will set up an inward radial flow of air along the sea surface. The ring is also a way to understand the origin of the eye of the hurricane, which is known to have a wide range of diameters, depending on the circumstances. This has the effect of increasing, overall, the rate of pumping moist air aloft.

A further change that lowers the pressure inside the imagined ring of thunderstorms, is the development of azimuthal air motion, air swirling inside the ring. The radially incoming air typically carries some angular momentum per unit mass, L. If this angular momentum is conserved, in the basic lack of friction in motion of large masses of air, then the local tangential velocity v will be expressed as:

$$V = L/r - fr/2, \tag{4.1}$$

where r is the radius from the center of the ring and f is the Coriolis factor, the angular momentum coming from the rotation of the earth at the latitude in question.

This equation shows a larger tangential speed at a small radius, related to the spin-up of the hurricane tangential winds, as the angular-momentum-conserving portion of the inflow moves to a smaller radius. It also allows, at a large radius, for the direction of the tangential motion to reverse, and this is observed at the top of the hurricane. The motion of the clouds at the top becomes anti-cyclonic, clockwise in the northern hemisphere. Incoming air at the sea surface, say in the first kilometer of altitude, does not fully conserve angular momentum, but above the frictional boundary layer (Eq. (4.1)) is likely to be observed.

The pressure gradient dp/dr needed to trap the circling air is described as:

$$dp/dr = \rho \, [v^2/r + fr]. \qquad (4.2)$$

This assumes a balance of radial centripetal acceleration and further pressure drop inside, again increasing the inward radial flow from the sea surface outside the ring. If the driving force is to humidify as large a portion of the tropical atmosphere as possible, two aspects come into view.

One is simply the rate per square meter of water being swept up as vapor, which is known to be increased by the speed of the air as it flows over the water surface.

A second factor is the volume of the atmosphere affected, which one can argue is enlarged by the hurricane formation to an area of radius of 600 km, or so, rather than the immediate area of the thunderstorms themselves. A countering tendency, however, is the return of much if not most of the water to the ocean in the form of hurricane-induced rainfall. So, it may be that an under-appreciated result of the hurricane is to pump initially-humid air into the stratosphere and away from the local ocean altogether. This flow preserves enthalpy and thus pumps enthalpy from the sea to the top of the atmosphere. The azimuthal wind aids the flow of water from the sea by lowering the internal pressure in the eye and by increasing the water release from the ocean per square meter by the larger local wind speed. However, most of the water thrown up into the troposphere by this process comes back to the ocean as rain.

Still, it may be the increase in the upper-level outflow of the sea level air preserving enthalpy that drives the whole process by decreasing the disequilibrium of humidity that was mentioned at the outset.

The conventional description of the hurricane is of a primary azimuthal airflow, the sea-level hurricane wind, and the secondary in-, up-, and out-flow that is more like the thunderstorm flow. In the model of Emanuel, the vertical flow is at constant enthalpy, and the upward flow is preceded by an inward radial flow near the sea surface in which the air greatly increases its enthalpy via its specific humidity.

Meteorologists often describe the increase of humidity in general terms, such as the increase in enthalpy, entropy, and equivalent potential temperature, but the operative change is the increase in water content of the air in grams per kilogram (g/kg), which controls the latent heat content. This secondary flow within the hurricane carries the conserved enthalpy,

$E = c_p T + gz + L_v q + 1/2v^2$, to the top of the troposphere or above.

It appears that the composition of the conserved enthalpy in an air packet at high altitudes includes less water, which has been precipitated out, and more kinetic energy, including an updraft of rapidly moving air. According to the model, the upward and outward flow is uniform and axisymmetric, but the satellite optical image usually shows a few favored directions and outflow jets, and these are actually rotating in the opposite sense to the cyclonic sense, counterclockwise that applies near the surface. The altitude of the top hurricane clouds is thought to be around 12–14 km. These are cirrus or cirrostratus clouds made up of ice crystals. The water vapor image of Fig. 1.3 is more accurately circular, the rain is concentrated at lower altitudes.

We return to the simple approximation of a hurricane as a ring of thunderstorms. Hurricanes come in different sizes. For the storms, shown on the cover of this book and Hurricane Isabel shown in Figs. 1.1 and 1.2, the scale is suggested by a 200 km radius for the region of peak winds, and a 600 km outer radius. If there were a circle of thunderstorms at, say, 0.75 of the radius of maximum winds (100 km), then about 72 thunderstorms would have been needed taking their radius as 3 km. Hurricane Harvey shown in Fig. 1.3 is smaller, with a clearly-defined, accurately-circular rain-free eye, with a radius of 20 km and an outer radius of about 193 km. For such a storm, the number of thunderstorms in the circular eyewall would be closer to 30. Figure 1.3, based on water vapor content, shows a circular structure with no suggestion of discrete thunderstorm constituents.

4.2 Atmospheric Ozone as a Hurricane Diagnostic

A different kind of image is shown in Fig. 4.1a (Zou and Wu, 2005).

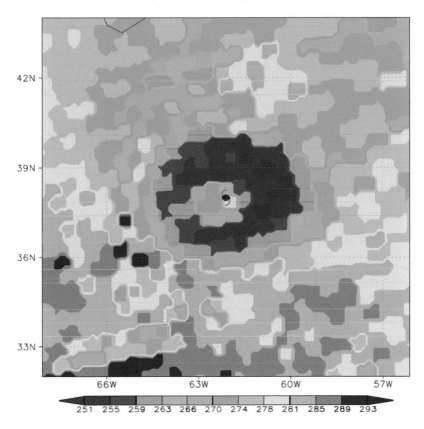

Figure 4.1a Image of ozone obtained above Hurricane Erin on September 12, 2001, using Total Ozone Mapping Spectrometer (TOMS) from NASA Earth Probe satellite. The color scale in Dobson units, such that 300 DU means 3 mm equivalent height of ozone column if compressed to standard conditions. The grid size is 3 degrees latitude, about 323 km, so the whole image has an overall dimension of 1292 km on a side. The blue area is of nearly uniform deficiency of ozone interpreted as nearly uniform upward airflow within the hurrican. Reproduced from Zou and Wu, 2005, with permission of John Wiley and Sons.

Figure 4.1b Visible image of Hurricane Erin on September 12, 2001. The white frame is the area shown in Fig. 4.1a of Erin using the ozone detector. This location is roughly northeast of New York City, and the Hurricane is becoming somewhat ill-defined, the eye region is not accurately circular. The white framed portion of the image has an overall dimension of 1292 km on a side. NOAA-12 AVHRR three-channel color composite of Hurricane Erin at 2031 UTC. Image provided courtesy of Steven Babin and Ray Sterner of Johns Hopkins University Applied Physics Laboratory. Reproduced from Zou and Wu, 2005, with permission of John Wiley and Sons.

Ozone (O_3) in the earth's atmosphere is a trace gas peaking in a concentration near an altitude of 23 km, in the stratosphere, where it is formed photochemically. However, about 10% of the total ozone is present in the troposphere, below 10 km, closer to the earth. Ozone is formed when oxygen atoms, released by optical absorption and splitting of oxygen molecules, join existing O_2 molecules. Ozone is important in absorbing ultraviolet light from the sun, which would be harmful if it reached the lower atmosphere.

In Fig. 4.1a, ozone is revealed, surprisingly, as a means of imaging a hurricane. The image obtained on September 12, 2001, shows, in yellow and green, a replica of the eye of Hurricane Erin, with a radius of around 30 km, and in blue, a wider annular region of a radius of about 161 km that appears to represent the top of the "eyewall cloud" of the hurricane. Zou and Wu stated that the blue (ozone deficient) region represents a "strong upward lifting of the atmosphere", locally blowing away upward the normal ozone, and replacing it with minimum-ozone air from the ocean surface. The orange (ozone-rich) center above the eye indicates "downward subsidence of the stratospheric atmosphere within the central region of the hurricane". The color image is based simply on the total amount of ozone in the air column below, calibrated in Dobson units, each representing 0.01 mm of ozone in the vertical column if compressed to standard conditions, STP. So, the changes in ozone amount in the image from 250 to 290 DU, correspond to removing most of the ozone from the body of the hurricane, extending up through the total troposphere. The red spots in the blue portion of the image are argued, by Zou and Lu, to be artifacts related to the reflection of light from local cloud tops. Counter to this argument, the red dots in the blue region might be regarded, according to the color scale, as minima in ozone, thus maxima in blowing sea-surface air from the hurricane beneath. The diameter of these red spots is about 0.15 grid units, thus 48.5 km. Their radius then is 24 km, about 8 times what we found in Chapter 3, namely 3 km for the radius of a typical thunderstorm. So, we suggest that these red-dot areas in Fig. 4.1a may be large thunderstorms that lie in the rain-producing region of the hurricane and are seen by their internal and overshooting updrafts that rearrange the ozone distribution. The whole blue region represents rising ozone-deficient air, and on the left side of the blue region of the image, the upcoming wind seems homogeneous, without distinct source centers, as one might expect for a smoothly circular hurricane image (Fig. 1.3) of Harvey at landfall.

This makes clear that the hurricane is a huge disturbance of the local atmosphere, whose influence and attendant flows of air reach the full height of the troposphere. It also appears to substantiate the view of the eyewall as arising from a ring of thunderstorms, and of thunderstorms as machines that produce huge updrafts moving ground-level air up even into the stratosphere. The interpretation is that the existing ozone layer is "advected" (carried along in the wind) and some of it goes back down to the surface in the region directly above the eye. This is direct evidence that stratospheric air flows down from above into the eye, carrying ozone with it. It is also known that the air above the eye is abnormally warm as the result of the release of latent heat by the condensation of water vapor flowing upward inside the eyewall clouds, which appear to be individually revealed in this image. An abnormal flow of dry warm air down to the eye, consistent with an abnormally low atmospheric pressure in the eye, helping the inward flow of sea level moistened air, is consistent with this picture.

An optical picture of Hurricane Erin is shown in Fig. 4.1b. It is not as uniform as the image on the cover of this book, and the eye region seems disrupted and there is a gap region of a slightly larger radius. This image may give support to the idea that Erin at this stage was decaying and showing discrete thunderstorm features as shown in Fig. 4.2.

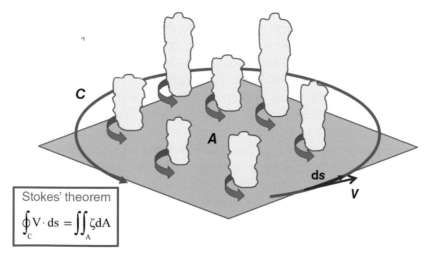

Stokes' theorem
$$\oint_C V \cdot ds = \iint_A \zeta dA$$

Figure 4.2 Sketch suggests an array of thunderstorms as a model of a hurricane. Zeta is the vorticity, the angular velocity about the normal direction. Reproduced from Smith and Montgomery, 2016, with permission of John Wiley and Sons.

The sketch in Fig. 4.2 suggests that a collection of locally rotating thunderstorms can act as the source of a circling azimuthal wind as in a forming hurricane. Mathematically, the area integral of the vertical component of the vorticity gives this result that depends on the definition of the vorticity zeta as the curl of the velocity vector. This sketch is offered by Smith and Montgomery as a situation that might arise in the formation of a hurricane, with the understanding that as the storm fully forms the ring of thunderstorms merges to form the featureless eyewall cloud. The ozone image in Fig. 4.1a seems to plausibly represent such a formative or more likely decaying stage for Hurricane Erin of September 12, 2001.

4.3   Consensus Structure of a Mature Hurricane

A summary of the mature hurricane in two panels is shown in Fig. 4.3 (Houze, 2011). The vertical coordinate is the altitude, presented as the pressure in mb, while the abscissa is the radius from the axis of azimuthal rotation. The left panel shows the flow of air from the lower left, the large radius at sea level, inward the eye and then upward to the top of the storm at 13.6 km, 150 mb pressure in this figure. Black arrows in the white eyewall cloud region suggest flows that are stronger near the eye. The red dashed lines are trajectories of flow with various values of equivalent potential temperature θ_e. Equivalent potential temperature θ_e is the temperature of a parcel of air. If the air adiabatically moved to the sea level, all of its water content q converted directly to heat via the relation $dq\, L_v = C_p\, dT$, with q the specific humidity. Along each red flow line, this number θ_e is constant. Larger values of this parameter, ranging up to 80 °C, indicate larger values of enthalpy, driven by water content. These red paths take enlarged enthalpy from the sea surface to the top of the clouds at 13.6 km. The strongest flows are those closest to the eye itself, at a small radius, and the arrows are directed nearly vertically. These upward flows, at their highest altitude, are constituted, as their water content is precipitated, more largely of kinetic energy and heat, rather than latent heat, as the air rises vertically and can provide "overshooting tops" as shown in the thunderstorm in Fig. 3.5.

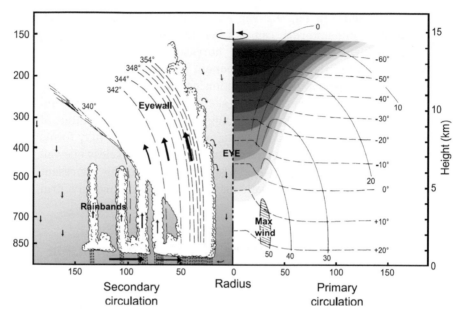

Figure 4.3 Summary of mature hurricane features: altitude in mb vs radius in km. Left panel: Red lines are loci of constant equivalent potential temperature, θ_e ranging up to 354 K or 81 °C. Equivalent potential temperature is the temperature of a parcel, which moved adiabatically to 100 kPa and if all its water contents were converted directly to heat. Left shows the secondary circulation of air: in, radially, along the sea surface, then vertically upward at the eyewall. Right panel: Primary azimuthal circulation. Reproduced from Houze, R., Jr., 2010. © American Meteorological Society. Used with permission.

This flow noted as secondary is actually the driving flow of the hurricane, fed by enthalpy release from the ocean. The primary, azimuthal flow of the hurricane wind is in some way fed by the secondary flow. The main features of the primary flow are shown in the right panel of Fig. 4.3. The region of high winds is shown at a low altitude, just above the frictional surface boundary layer. The dashed and solid contour lines, respectively, in the right panel, show temperature in Celsius and azimuthal wind speed in meters per second. The highest azimuthal wind speed is located at the eyewall, and its largest contour is labeled 50 m/s. Beyond this point, wind speeds fall off in both the radial and vertical directions. In addition, the azimuthal wind decreases as one moves down from the height of optimum speed, toward the sea surface. The air drags on the sea to induce wave action and also slow the wind speed. The dragging of the wind along the foamy sea surface is also the place where the essential increase of equivalent potential temperature occurs, the moistening of the air, which is then pulled aloft by deep convection as in a thunderstorm, in the eyewall cloud. The temperature contours are consistent with ≈28 °C at the ocean surface. Note at the top of the right panel of Fig. 4.3 is shown a solid contour line of zero azimuthal speed. At higher altitudes, the azimuthal flow reverses to the anti-cyclonic sense. The color scale is of potential temperature vs altitude and radius. The hurricane is a warm core situation, the air above the eye itself is anomalously warm by more than 10 °C and this helps make the sea-level air pressure at the storm core very low, to drive the radially inward flow of moist air along the sea surface.

The left side of Fig. 4.3, the secondary flow of a hurricane, has the essential features of a thunderstorm, drawing in air at the bottom, the air rises, floating upward by the deep convection fed by the water content of the rising parcels. This provides a steady flow of enthalpy from the sea to the upper atmosphere, an improvement over the thunderstorm that is only transient. The flow of moist air vertically carries conserved enthalpy to the stratosphere, where the parcels contain enthalpy less in the form of latent heat and more in the form of temperature and kinetic energy. This warms the full height of the eye region, (thus lowering the air pressure) and the kinetic energy provides the "overshooting top" phenomenon.

This enthalpy flow is enabled by the thunderstorm, and in an enhanced fashion by the hurricane. According to Emanuel, the disequilibrium corrected by the flow we are discussing is the driving force behind hurricanes. The flow shown in the left panel would not exist without the hurricane. The structure of the hurricane enhances the upward enthalpy flow over that provided by a thunderstorm, i.e., smaller, ≈ 3 km in radius, and transient. The hurricane serves to stabilize the flow and multiply the flow by its scale on the order of a dozen or more thunderstorms forming its inner eyewall cloud structure. The right panel shows the azimuthal flow, but gives no clue as to how it forms.

Speaking roughly, it appears that the hurricane azimuthal flow increases the rate of water release to the atmosphere from the ocean, because the air speed roughens the surface, increasing its area in addition to simply bringing more air per unit time, to each square meter of the ocean surface. At the same time, the azimuthal flow increases the pressure drop toward the center of the need to provide dynamical balance for the centripetal force. The greater pressure deficit at the origin increases the flow of air inward and thus the flow of enthalpy to the stratosphere. It appears that the hurricane's azimuthal flow is acting as a catalyst to increase the overall enthalpy flow. The azimuthal motion is parasitic upon the secondary flow but allows the scale of the secondary flow overall to be enlarged, directly addressing the noted sea-atmosphere disequilibrium. The hurricane azimuthal flow is not directly lowering the energy of the ocean-atmosphere system but is a collateral feature that enhances the overall flow, as shown on the left side of Fig. 4.3.

It appears that the mechanism for redirecting a portion of the convective energy into the azimuthal motion is not well understood. A key question is how the angular momentum is provided. The literature often mentions that the "ambient vorticity", the tendency of the atmosphere, in response to the rotation of the earth, to be spinning in a counterclockwise sense, is collected by the inward radial airflow and concentrated in the eyewall cloud, helping it to spin. The need for initial rotation to start the hurricane is an argument for this view, but the question seems to remain whether the ambient vorticity is sufficient.

One can speculate that the spiral motion of the incoming air at the sea level increases the enthalpy flow by lengthening the path along the tops of the waves, increasing the specific humidity of the air that is convected upward. The spiral motion, lengthening the path for moisturizing, increases the desired flow. The system may find some way to foster that spiral flow by creating vertical angular momentum. This is a complex subject but it does connect to an added area of knowledge related to rotating storms and tornadoes, alluded to in connection with Fig. 3.5.

One general idea is that wind shear creates angular momentum, rotation, about a horizontal axis, and that rotation may be redirected vertically by an updraft as occurs at the center of the convective cell of the thunderstorm or the center of the hurricane. A radial pattern of wind shear exists around the hurricane due to the more rapid air in-flow to the center at the ocean surface vs at a higher altitude.

Figure 4.4 emphasizes the boundary layer in a model of the inner core region of a hurricane. The boundary layer demarked by dashed lines is between 0.5 km and 1 km in depth. Air falls down, subsides, into boundary layer for radii larger than marked r_{up} and rises up, ascends, from boundary layer at radii smaller than r_{up}. The near sea level inflow, driven by a frictionally induced pressure gradient, culminates in an inward radially flowing jet of moist air, which begins to rise at the marked radius r_{up}.

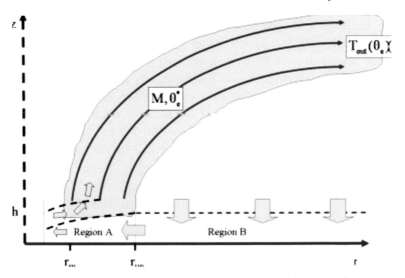

Figure 4.4 Detail of the boundary layer in the model of the hurricane. Reproduced from Smith et al., 2009, with permission of John Wiley and Sons.

The frictionally induced convergence, or inflow, in the boundary layer, is thought to force deep convection, by providing the supply of moist air, which generates a larger inflow through a tall tropospheric layer lying above. The inflow also brings in angular momentum and confers some of that angular momentum to the tangential cyclone wind field that is most intense near this radius and at the top of the boundary layer.

4.3.1 Energy Estimates on a Carnot Engine Model

Numerical estimates for Hurricane Isabel, depicted in Figs. 1.1 and 1.2, with an overall radius of 600 km, were offered in Chapter 1. These are very approximate. The power in the azimuthal wind was estimated as 150 TW, and the same number was arrived at for the dissipation of power into the ocean, again 150 TW. The overall power corresponding to the rainfall was estimated as 441 TW. If this total power is the sum of the azimuthal wind power and the vertical updraft wind power, which is exhausted to the stratosphere as shown in Fig. 4.1a, then that updraft power, the enthalpy flow to correct the disequilibrium between sea and atmosphere for the Isabel sized (large) hurricane is 291 TW.

Emanuel and others have discussed the hurricane as a heat engine following the Carnot cycle. The Carnot engine does work W by taking in heat Q_H from a hot reservoir and returning heat Q_C to a cold reservoir. The energy balance is $W = Q_H - Q_C$ and the efficiency $\eta = 1 - T_C/T_H$. The efficiency is taken as 1/3 as a result of choosing T_C = 200 K and T_H = 300 K. Since we have estimated W = 150 TW, we find simply that Q_H = 450 TW and Q_C = 300 TW. This latter value is to be compared with 291 TW found above, in reasonable agreement. So, the driving force to reduce disequilibrium is represented by the exhaust flow of the Carnot cycle, about 300 TW for our estimate of Hurricane Isabel. By making use of the rainfall numbers, we got a similar answer without invoking the Carnot model for the hurricane.

As a numerical test, we can ask what exit wind speed would correspond to the estimated 300 TW at the top of Hurricane Erin as depicted in Fig. 4.1a. The blue area in that figure is an area of large upward airflow according to the authors of the paper. A uniform upward flow of air at speed v crossing area A carries a power $P = A\rho v^3/2$. From the dimensions of the blue area in Fig. 4.1a, the blue annular region has an inner radius of 30 km and an outer radius of 161 km, and thus area A = 78.6 Gm². Based on Eq. (2.8), we take the mass density at 13.6 km altitude as 0.43 kg/m³. Using the power equation $P = A\rho v^3/2$, setting P to 300 TW we find v = 20.0 m/s for the speed of the air in the blue region of Fig. 4.1a. This is a reasonable answer, although we have simply assumed that the exhaust power of Hurricane Erin is similar to that of Isabel.

So, we may say that the azimuthal flow of the hurricane stabilizes and enhances a ring of thunderstorms, in the form of the eyewall cloud, with the end result of sending a high-speed overshoot of cold air into the stratosphere, carrying power (enthalpy per unit time) on the order of 300 TW for an Isabel sized storm, reducing the noted enthalpy disequilibrium between tropical ocean and atmosphere. It has been remarked by Emanuel (2018) and references therein that the hurricane outflow or overshoot has been observed by high flying aircraft to be generally turbulent, consistent with a high-velocity output of cold air. The turbulent nature of the outflow was adopted as a key point in the theoretical modeling of Emanuel and Rotunno, 2011.

4.4 Detailed Look at Hurricane Rita (2005)

Data summarized in the sketches of Fig. 4.3 have been accumulated from research aircraft flying into the storms. Striking photographs have been obtained as the airplanes have flown into the eye regions, penetrating the eyewall at some intermediate altitude, perhaps 5 km. What is seen in some cases is blue sky overhead, with an accurately circular array of rising clouds on all sides representing the eyewall. The appearance is of being inside a huge circular football stadium, with the diagonally oriented rising eyewall looking like the bleacher seats, rising up to 10 km. One of the charming aspects of the meteorological literature is that a diagonally rising cloud-like eyewall is said to be "slantwise" and the angular momentum brought in by the inward flow of sea surface air is said to be "advected" vertically, advection being the process of being carried along in the wind.

The tools available on the research aircraft include radar, particularly dual Doppler radar. Radar is the centimeter wavelength radiation that reflects back from liquid water but easily penetrates clouds. The wavelength shift of the reflected microwave, if the reflecting surface is moving, is the Doppler shift that is recorded. So, the relative motion of the reflecting rain is measured by these research aircraft. If the radar image in the Doppler radar is split, meaning some of the reflected radiations are shifted up and some are shifted down in wavelength, then rotation as in a tornado can be detected by the Doppler radar.

An interesting study of Hurricane Rita of 2005 is shown in Figs. 4.5a,b (Houze, 2010). The color scale in the first image, reconstructed to show a vertical view of the hurricane, is simply the radar reflectivity, directly related to the liquid water content. So, the red annular regions, about 5 km wide, are filled with rain and are interpreted as eyewall clouds. The interesting feature is that there are two having an equivalent size in this episode of the life of Hurricane Rita. Between the two eyewall clouds is a rain-free region whose properties resemble the eye itself, and this region is called the moat. Hurricanes quite frequently adjust their overall size by changing their eyewall clouds and this seems to happen by the growth of the second eyewall cloud at a larger radius, by the coalescence of rain bands that are somewhat like circles of thunderstorms, into a second eyewall cloud. Eventually, the inner eyewall reduces in radius and goes away and the outer eyewall takes its place.

Figure 4.5a shows a radar reflectivity image obtained by aircraft flying through Hurricane Rita of 2005 during an unusual period in which it contained two concentric eyewall clouds. The red circles (annular regions) indicate liquid water content (rainfall), and the authors mentioned that the two circles also correspond to two separate maxima in the hurricane wind speed, the azimuthal wind. The radii in this image are about 45 km, 25 km, and 18 km, respectively, for the outer eyewall cloud, the inner eyewall cloud, and the eye itself. So, this hurricane is much smaller than Hurricane Isabel of Fig. 1.1, where the eyewall cloud radius was on the order of 200 km. As mentioned, the region between the two eyewalls, the moat, acts as the eye region, with no precipitation falling. This image is based on liquid water, differing from the image of Fig. 1.3, which is based on total water including water vapor, referred to as total precipitable water. This storm is similar to Hurricane Harvey shown in Fig. 1.3 where the eye radius was 20 km, compared to about 18 km in Fig. 4.5a.

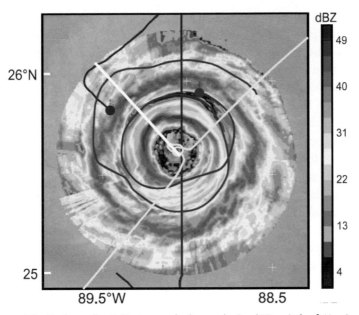

Figure 4.5a Radar reflectivity image (color scale in dBZ units) of Hurricane Rita of 2005 showing two simultaneous eyewall clouds. This image was obtained by aircraft flying through the storm on tracks shown in the image. The unusual double eyewall is a step in an eyewall replacement process. Reproduced from Houze et al., 2007, with permission of the American Association for the Advancement of Science.

The direct aircraft measurements on Hurricane Rita also included Doppler radar, from which the vector velocity of the reflecting water can be deduced. Figure 4.5b shows a cross-section image constructed for Hurricane Rita under the same conditions as in Fig. 4.5a. The color scale in this image is based on radar reflecting liquid water. The red regions on the lower right match with the concentric circles shown in Fig. 4.5a occurring at radii around 25 km and 45 km.

Looking at the higher altitude region above a radius of 25 km, the inner eyewall cloud, we see the vertical flow of air marked by the vector velocity arrows. Note the calibration wind velocity arrow of 10 m/s on the lower right. The red region above a radius of 25 km peaks around a height of 5 km, but extends from the ocean surface up to about 8 km. In agreement with Figs. 1.2 and 1.3, the rainfall is maximum in the eyewall cloud itself. The black wind vector arrows are primarily vertical in this whole region extending up to 12 km, the top of the yellow region where the wind arrows become longer and begin to turn toward an outward radial direction. The longest arrows appear to represent wind speeds up to 50 m/s, at an altitude of about 13.6 km. This resembles the overshooting wind phenomenon noted earlier. The diminishing liquid water content (red color scale turning to yellow color), as one moves up inside the inner eyewall cloud, coincides with increasing vertical wind speed as predicted in the analysis of convection in Eq. (2.14).

This image (Fig. 4.5b) allows an estimate of the wind power flowing out the top of the inner eyewall cloud of Hurricane Rita during its process of eyewall change. Consider the region in the vicinity of a radius of 25 km and a height of 13.6 km, taking the wind speed as 50 m/s. The wind power density is given by $P = 1/2\ \rho v^3$, where the air density is taken at that altitude as 0.43 kg/m^3. If the region is an annulus of width 5 km at a radius of 30 km (from Fig. 4.5), with an area of 10^9 m^2, we find the exhaust power as 26.8 TW.

We can get more information by assuming the hurricane acts like a Carnot engine, where the 26.8 TW plays the role of Q_C, the heat rejected to the cold reservoir. Recall that the Carnot engine does work $W = Q_H\ \eta$ by taking in heat Q_H from a hot reservoir and returning heat Q_C to a cold reservoir. The energy balance is $W = Q_H\ \eta = Q_H - Q_C$ and the efficiency $\eta = 1 - T_C/T_H$. The efficiency is taken as 1/3 as a result of choosing $T_C = 200$ K and $T_H = 300$ K. Since we have estimated $Q_C = 26.8$ TW, we find simply $Q_H = 40.2$ TW and $W = 13.4$ TW. This final number for W is directly comparable to the dissipation power described by Emanuel (see Eq. (1.6)) and falls within the range of 3–30 TW given by Emanuel, 1999. In this interpretation, the work done by the Carnot cycle is the source of the mechanical energy of the wind, and all of that mechanical energy is finally transformed into wave motion on the sea underneath the storm, called dissipation by Emanuel. The Carnot cycle repeats and thus the work performed per cycle is equivalent to the power expended by the hurricane interpreted as a Carnot engine.

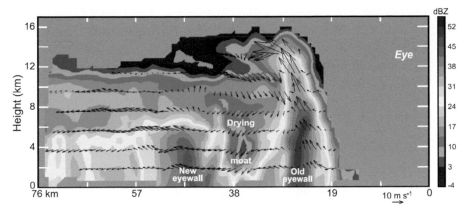

Figure 4.5b Radar reflectivity cross-sectional image (color scale in dBZ units) of Hurricane Rita of 2005 showing two simultaneous eyewall clouds. Vector arrows in this image represent local wind speed as deduced from aircraft Doppler radar, and a calibration vector of 10 m/s is shown on the lower right below a radius marker of 19 km. This image was obtained by NRL Naval Research Laboratory P-3 aircraft using NCAR Electra Doppler Radar (ELDORA) flying through the storm on tracks as shown in Fig. 4.5a. The unusual double eyewall is a step in an eyewall replacement process. Reproduced from Houze et al., 2007, with permission of the American Association for the Advancement of Science.

4.5 Ice Particles in Thunderstorms and Eyewall Clouds

The freezing point of water is reached in thunderstorms and eyewall clouds at altitudes in the vicinity of 5 km. The tropical ocean surface may be nearly 30 °C, so, at a nominal lapse rate of 10 K/km, –10 °C would occur at 4 km altitude. The cumulonimbus and eyewall clouds go up to 13 km, so most of these clouds are frozen (the word in meteorology is "glaciated"). There are a variety of frozen particles in tall clouds: snowflakes, graupel, hail, sleet, rime, aggregates, ice splinters, columns, and frozen droplets, with many names in use. Many of these are shown in Fig. 4.6. In this figure, the freezing point of 0 °C is shown near 5 km altitude, and the mixed-phase range is estimated as ±10 °C.

Most of these particles melt and finally fall down as rain, in the tropical storms at least. But large hails are known in warm places, but not typically under hurricanes. The tops of these eyewall and thunderstorm clouds seen from space are mostly cirrus, made of small ice crystals. While basically only liquid rain is found under the hurricane, a large fraction of that water had been frozen prior to its fall. As mentioned, large thunderstorms are known to have hail, even large-diameter hail, in addition to the rainwater underneath.

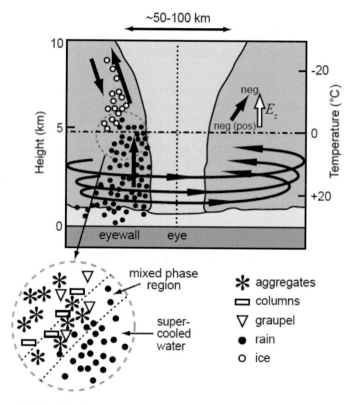

Figure 4.6 Sketch of precipitation particles and processes in a tall cloud near the freezing point in the atmosphere, typically 5 km. In this figure, open symbols are solids based on ice and dark symbols are liquid water. The word graupel refers to frozen aggregate particles big enough to usually be falling downward, sometimes called soft hail. It is suggested that graupel may result from a snowflake absorbing a supercooled liquid drop and freezing. As particles rise or fall through the 0 °C line they often overshoot before they change phase, the presence of supercooled liquid above the 0 °C line is an important factor. Reproduced from Houze, 2010. © American Meteorological Society. Used with permission.

A separate fact is that the electrification of clouds allows lightning to produce thunder, which occurs only when ice particles are present. Collisions between ice crystals are needed for the transfer of charge. Liquid drop collisions or liquid-solid collisions never produce charge transfer, nor thunder. So, a thunderclap means an ice-containing cloud of the order of at least 5 km height is involved. Typically, such ice-ice collisions are between a large falling ice particle, perhaps referred to as graupel or an aggregate, against a smaller rising splinter of ice, lifted by the typical updraft of these clouds perhaps 5 m/s at the altitude of the mixed phases where liquid drops and ice particles are both present.

The lightning strike means that ice particles of negative charge have accumulated in a region separate from the partner particles of opposite charge in sufficient numbers to create an electric field of megavolts per meter to a create breakdown of the air. This certainly happens in thunderstorms but has been found to be rare in eyewall clouds. So, it may be that the eyewall cloud allows the small negatively charged particles to keep rising far upward in the strong updraft and not to congregate as they seem to be in the thunderstorm, to allow a lightning strike. Measurements show that lightning from the eyewall cloud is almost non-existent while lightning from clouds in outer rain bands of the hurricane is quite common.

One of the puzzling aspects of ice particles in tall clouds is their large number. It is widely reported that the number density of ice particles in the vicinity of −10 °C, is up to three orders of magnitude higher than is reasonably expected. Indeed, cloud microphysical data obtained by aircraft in tropical cyclones show the storms to be great producers of ice in the mid to upper troposphere. The schematic Fig. 4.3 shows that the hurricane is a massive fountain of upper-level cloud, which is nearly all glaciated. The eyewalls and rainbands produce large amounts of rain by the coalescence of drops. The eyewall clouds also generate large amounts of graupel, sometimes called soft hail. The "riming" process (growth of an ice particle by absorbing supercooled liquid drops) produces tiny ice splinters by the Hallett-Mossop mechanism (Hallett and Mossop, 1974). The tiny splinters are lofted up and spread widely around the hurricane, eventually contributing to weak rainfall over a wide area as the splinters finally fall down and melt.

To form an ice particle, in practice, a nucleating center must be present. The nucleating center allows water molecules to congregate close to one another and form a frozen particle (the process is similar to forming a liquid drop of rain at a higher temperature). The ice-nucleating particles are thought to be aerosols small enough to remain aloft indefinitely. Some well-known aerosols are based on sulfur dioxide and on nitrogen compounds. The informed opinion is that the number of such nucleating sites for atmospheric ice is too small to explain the measured density of ice particles in tall clouds, and the discrepancy is up to three orders of magnitude.

This has led to the widespread belief that there is at least one potential mechanism of secondary ice production. In some way, the density of small ice particles in glaciated clouds must be multiplied by collisions among primary ice particles. The leading candidate is the Hallett-Mossop process (Hallett and Mossop, 1974), which is a collision of a supercooled liquid drop with a large cold ice particle (Fig. 4.7).

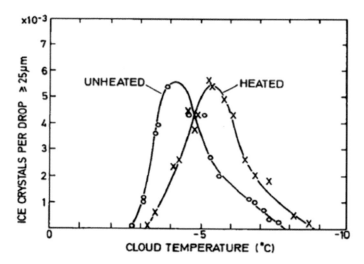

Figure 4.7 Ice crystals per drop: Production of ice splinters per encounter with supercooled water droplet of 25-micrometer size. Laboratory modeling of riming, the growth of ice particles by accreting water droplets, with a secondary production of hundreds of splinters. Reproduced from Heymsfield and Mossop, 1984, with permission of John Wiley and Sons.

In some way, chaotic freezing of the supercooled water drops, on the order of 15-micrometer size, aided by the energy release of the latent heat of fusion, produces a spray of tiny ice splinters. This process has repeatedly been demonstrated in laboratory cloud chamber experiments and shows a peak in the production rate when the temperature of the supercooled drop is in the vicinity of –5 °C.

4.6 Large-Scale Atmospheric Effects on Storm Formation

The sea surface temperature is the single parameter that controls the strength of hurricanes as shown in Fig. 1.6 for the power dissipation index (PDI).

At the same time, there are additional features of the earth's atmosphere and ocean that influence the probability of forming a hurricane. These factors seem to average out in considering the PDI, which is integral over a year's time and over a whole ocean basin like the North Atlantic. In large part, the extra factors that are important in predicting storms influence the resulting hurricanes only by their influence on the sea temperature. The additional factors include aerosols in the atmosphere, wind shear above the site, the el Nino and la Nina effects on the oceans, and the Julian-Madden oscillation. The latter is a globe-encircling vertical oscillation in the atmosphere. This is a wave of vertical atmospheric motion that slowly propagates around the equator. When the upward portion of the Julian-Madden wave is over the North Atlantic, it will assist the vertical flow that is a central feature, along with the related azimuthal wind, of the hurricane. To include some of these effects, the basic picture of the atmosphere as described in Chapter 2 needs to be expanded to include additional features that clearly have an influence on the probability of the formation of a hurricane.

An extreme example of such an atmospheric factor is shown in Fig. 4.8.

Figure 4.8 is an image of a dust cloud extending from the Sahara Desert in Africa toward the Yucatan Peninsula in Central America, a distance of some 4000 miles. The dust resulted from a windstorm in the Sahara, blowing tiny rock fragments and dust particles, up into the atmosphere where they are evidently transported by high-altitude winds in a westward direction across the Atlantic. These particles must be in the aerosol size range to remain aloft for a week's time needed, see Eq. (2.17). They are seen in Fig. 4.8 by their infrared radiation upward to the satellite's camera. The particles are in an altitude range of 1–2.5 miles, as indicated by laser image detection and ranging (LIDAR) measurements at the University of Miami.

Figure 4.8 Satellite view of 4000-mile-wide dust cloud emanating from the Sahara Desert in June 2020. See Florida in the upper left and Yucatan Peninsula in the mid-left of view. The color scale indicates radiation from dust particles in the aerosol size range that crosses the Atlantic Ocean before depositing (Cappucci, 2020c). https://www.washingtonpost.com/weather/2020/06/23/saharan-dust-hurricanes/.

There are several effects of this huge and unusual aerosol cloud (Cappucci, 2020c).

The first effect is to reduce the sunlight on the ocean below by reflecting some sunlight back upward, to decrease the ocean temperature. In practice, the dust inhibits the formation of clouds in fact often makes the local weather warmer, an opposite effect.

The second effect is to locally heat the atmosphere, as the particles intercept sunlight and thereby warm themselves and the nearby air. A warm layer in the atmosphere above the ocean will stop convective updrafts, as mentioned earlier, which depend upon a warm packet rising in a cooler environment, see Eq. (2.14). So, the dust cloud will stop the formation of thunderclouds, which are needed to start a hurricane.

Finally, the whole mass of air coming from the Sahara Desert is inherently dry, lowering the humidity above the ocean where the cloud passes. These effects of the unusually strong aerosol (Fig. 4.8) are all deleterious to starting a hurricane.

On the other hand, the dust cloud aerosol phenomenon of Fig. 4.8 is temporary. It surely lowers the chance of any hurricane soon, but over a year's time, it probably will have little effect on the PDI, because of its short duration. The importance of this recent aerosol is its great reduction in the near-term chance of a hurricane, an important aspect of hurricane forecasting. To forecast the long-term damage done by hurricanes, the overall PDI is shown to depend only on sea temperature by Figs. 1.5 and 1.6, but in the short-term prediction of a storm, other factors such as the dust cloud need to be looked at closely.

For a long-term aerosol effect on PDI, we can look back at the original data of Emanuel, as shown in Fig. 1.5, which shows the North Atlantic PDI and offset ocean surface temperature vs time over recent decades. The large increase, near doubling, in both PDI and the offset temperature $(T - Tc)$ since 1980 is believed to be related to the reduction in a continuous aerosol cover of the North Atlantic, as particulate emissions from coal-burning electric power plants in Europe were reduced over a period of decades. As shown in Fig. 1.6, the aerosols were carried westward across the Atlantic by high-level winds. So, reducing the particulates warmed the North Atlantic and increased the PDI. The unrecognized aspect of the data was that the near identity of the PDI and offset temperature curves indicated that the aerosol's effect on the PDI was only through its effect on the ocean temperature.

Another condition that inhibits hurricane formation is wind shear, i.e., a difference in the wind at altitude vs at sea level. This tends to disrupt the formation of a hurricane. The discussion above for the North Atlantic shows that there is a westward wind that carries dust particles from the Sahara Desert to the Americas, as well as carrying particulate aerosol from Europe out over the Atlantic Ocean. In general terms, hurricanes tend to move along with the local air mass, from east to west. In addition, dry conditions in the atmosphere, as with the Saharan air, make it harder for a hurricane to form.

4.7 Potential Intensity, a Triumph of Computational Meteorology

The potential intensity, in meters per second, is an estimate of the maximum sea level wind speed V that could result from a hurricane if it was to form under the conditions at hand. These estimates have been attempted repeatedly in the historical meteorology literature, and in recent decades, these have been notably advanced by Prof. Kerry Emanuel, starting from the first principle approach based on the Carnot cycle (Emanuel, 1986, 1991); see Eq. (1.13).

The thermodynamical potential intensity, calculated using modern computational tools, and shown in Fig. 4.9, has been nearly universally adopted by the meteorology community, even to the point of calculating PDI values based on this quantity V, a theoretical construct.

Following Bister and Emanuel (2002) and Emanuel (2018), a slightly altered version of Eq. (1.13) is used to start the modern potential intensity calculation:

$$V_m^2 = (C_D/C_k)\,(T_s - T_o)/T_s\,(k_o^* - k_a) \qquad (4.3)$$

The only changes from Eq. (1.13) are replacing T_t (top) with T_o (outflow) and inserting the ratio of coefficients C, since the change in the final term is simply a change in notation from enthalpy E to enthalpy k, with units J/kg. The enthalpy k_a is that at a nominal altitude of 10 m in the ambient boundary layer, as compared to the saturation value k_o^* right at the water surface: $(k_o^* - k_a)$ is the fundamental enthalpy disequilibrium driving the storm. The ratio of drag coefficient C_D (see Eq. (1.6)) to enthalpy transfer coefficient C_k at the ocean surface is of order unity, each coefficient on the order of 10^{-3}. The basis of the expression is interpreting the airflow inward and upward (and return) in the secondary flow of the hurricane (left panel of Fig. 4.3) as a Carnot cycle (Emanuel, 1986) as described in text near Eq. (1.13). The basic idea is approximating the secondary flow as a rectangular trajectory in a temperature-entropy diagram, the area enclosed by the two isothermal and two constant-entropy curves representing the work done, which showed as V_m^2 in Eq. (4.3).

Equation (4.3) (Emanuel, 2018) has at least four unknowns: V_m, T_o, k_o^*, and k_a, all values applying at the eyewall of the presumed hurricane. It is the genius of Bister and Emanuel to have found credible approximate iterative solutions to this equation including an iterative algorithm that led to Fig. 4.9. At each pixel in this image, the solution for V has utilized satellite data including sea surface temperature, and a profile of air temperature, air pressure, and air humidity from sea level to the top of the troposphere. In their iterative approximate solution, it was further assumed, in calculating the boundary layer enthalpy k_a, that the relative humidity of the boundary layer at the radius of maximum winds is the same as that of the unperturbed environment.

According to Emanuel (2018), in this expression, the k_a boundary layer enthalpy is not known, nor is the surface pressure needed to calculate the surface enthalpy k_o^*, nor the outflow temperature T_o. The pressure dependence of the surface k_o^* together with the decrease in temperature that occurs as inflowing air attempts to cool adiabatically as it flows down a pressure gradient and increases the enthalpy disequilibrium, $k_o^* - k_a$. This is positive feedback, as increasing V yields a greater pressure drop.

A related but different aspect of the surface pressure drop δp coming into the eyewall from a large radius is the empirically verified expression:

$$\Delta p = -2.5 \, \Delta\theta_e \qquad (4.4)$$

in millibars, where θ_e is the equivalent potential temperature. θ_e is the temperature that results at standard conditions if all the water vapor content in a parcel is turned into heat, so a rise in θ_e will occur isothermally with a rise in humidity. θ_e is seen to increase as the radius comes into the eyewall in the red curves in the left panel of Fig. 4.3. This is an indication of the increase in enthalpy of the air, as it approaches saturation coming toward the eye at the ocean surface level, along with the first, nominally isothermal, leg of the Carnot cycle.

Returning to the Bister and Emanuel (2002) solution of Eq. (4.3), the authors specify the location of their evaluation to be at the eyewall, the radius of maximum winds, and make use of a thermodynamic identity to express V^2 as:

$$V_m^2 = (C_D/C_k)\,(T_s/T_o)\,[CAPE^* - CAPE]_m \tag{4.5}$$

(at the radius of maximum winds, R_m). Here CAPE is the convectively available potential energy, defined above in Eq. (2.13). The lower limit of integration for CAPE* is the sea surface at the radius of maximum winds, and for CAPE, the integration starts at the ambient boundary layer at the nominal altitude of 10 m. The researchers have used available satellite data to find the temperatures, humidities, and pressures for altitudes from sea level to the tropopause, which are needed for the calculation.

An additional assumption related to Eq. (4.2), for the pressure gradient to balance the centrifugal forces, is made, to use Eq. (4.5). This leads to the second operating equation:

$$c_p\,T_s\,\ln(p_o/p_a) = [1/2\,V_m^2 + CAPE]_m, \tag{4.6}$$

where c_p is the specific heat at constant pressure and the pressures p is evaluated at the sea surface p_o and the boundary layer p_a. (The pressure p_a at the sea surface is one of the defining measures of hurricane strength, as mentioned in Chapter 1.) The added assumption of Eq. (4.6) has been criticized by Smith et al., 2009, whose comments might lead to slightly different potential intensity values. Following Bister and Emanuel (2002), the two operating equations (Eqs. (4.5 and 4.6)) taking values of C_D and C_k, the value of T_s, and the ambient profile of temperature, become closed relations to determine V_m and p_o. CAPE is calculated by a parcel lifting algorithm, a computer program. It appears that the authors have shared their program with colleagues, allowing this elaborate procedure to be widely used in the meteorology community. Owing to the pressure p dependence of CAPE* and CAPE, the two equations (Eqs. (4.5 and 4.6)) must be solved iteratively using the satellite data sets.

The output of this "first principle" but the very complicated and approximate computational process is shown in Fig. 4.9, with color scale values of potential intensity V running up to 90 m/s. As stated by Emanuel (2018), intensity values from the theoretical construct have a basic similarity to the observed distribution of hurricane speeds, peaking in the tropics and fading away at high latitudes.

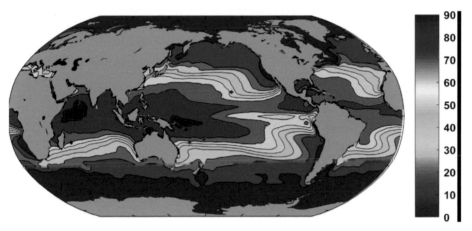

Figure 4.9 Calculated annual maximum of the potential intensity V in m/s based on satellite data averaged over 1979–2016 with a spatial resolution of 1.5 degrees, using the algorithm of Bister and Emanuel, 2002. Data at each pixel includes sea surface temperature and air temperature, humidity, and pressure from the sea surface to the top of the troposphere. Reproduced from Emanuel, 2018. © American Meteorological Society. Used with permission.

One clear discrepancy in the procedure is to predict hurricane winds at the equator, where in fact they hardly ever occur, and this failure results simply from the assumption in the Carnot cycle derivation that a vortex circulation is initially present everywhere.

Again, following Emanuel (2018), the assumption, in solving Eqs. (4.5 and 4.6) (that replace Eq. (4.4)), is made that the relative humidity at the eyewall boundary layer is the same as that of the unperturbed environment. Given this assumption, the pressure effects on the surface saturation enthalpy k_o^* are calculated iteratively; once an initial estimate is made for the potential intensity V. This is used to estimate the surface pressure under the eyewall, which, in turn, is used to revise the initial estimate of the saturation enthalpy k_o^*. The iteration is said to converge using the available datasets.

The authors have remarked in applying their procedures with larger values of V are obtained when the sea surface temperature is raised. However, the scheme does not allow easy determination of the actual temperature dependence of the V. This is a theoretical construct, whereas the earlier data-based work of Emanuel (2005), as interpreted by Wolf (2020) reveals an experimental determination of the temperature dependence of V as:

$$V = \text{const.} \ (T - T_C)^{1/3}, \tag{4.7}$$

where the value T_C was found as 26.5 °C for the North Atlantic.

It would seem that the basic procedure of Bister and Emanuel (2002) could be adapted to look for the observed temperature dependence and critical temperature. One question is whether the procedure shows any solution below the experimentally observed critical sea surface temperature of 26.5 °C, consistent with the finding of Dare and McBride (2011) that hurricanes do not form below a sea surface temperature of 26.5 °C.

Chapter 5

Critical Aspects of Rainfall and Hurricanes

"Critical aspects" of rainfall and hurricanes are those aspects related to a second-order phase transition. Broadly speaking, a "phase" is a general organization of matter, for example as a solid or a liquid. The most familiar critical aspects are a transition temperature and critical exponent, occurring in a typical power-law relation such as Eq. (1.18). We (Wolf, 2020) have identified a critical temperature for hurricanes, as described in Chapter 1. That paper estimated the transition temperature as 26.5 °C and the critical exponent related to the wind speed as near 1/3. An excellent review entitled "100 Years of Progress in Tropical Cyclone Research" (2018) by Prof. Kerry Emanuel does not include the word "exponent" and has no discussion of a hurricane as possessing a critical temperature.

"Critical phenomena" have been intensely studied only more recently with the discovery of subtle mathematical relationships leading to the Nobel Prize in Physics in 1982 to Kenneth G. Wilson. The overlap of this field with meteorology has been limited to the discovery of critical aspects of rainfall and Wolf's (2020) discovery regarding hurricanes.

Physics and Future of Hurricanes
Edward L. Wolf
Copyright © 2023 Jenny Stanford Publishing Pte. Ltd.
ISBN 978-981-4968-54-6 (Hardcover), 978-1-003-33125-4 (eBook)
www.jennystanford.com

5.1 Phase Transitions

The familiar transitions between solid, liquid, and gas phases of matter are the first-order transitions, with discontinuous changes in properties such as density at the transition temperature, and requiring a latent heat for the transition to occur. The vaporization of water requires 540 cal/g as mentioned above. In contrast, the critical phenomena related to rainfall and hurricanes are continuous phase transitions, of the second order, with no latent heat needed.

5.1.1 Second-Order Phase Transitions, Ferromagnetism

Ferromagnetism is perhaps the most familiar example of a continuous phase transition. Iron metal spontaneously becomes magnetic when cooled below 770 °C, so it will act like a bar magnet. The origin is an internal change in the organization of the electron motions in the iron, and ferromagnetism can be described as a new state where all the electron spin moments are aligned. The new state is characterized by its magnetization M, the order parameter, and the density of magnetic moments per unit volume. This quantity appears continuously at T_c and rises rapidly below T_c. The new state is a state of "broken symmetry," because a new direction, that of the locked-together spins, has spontaneously appeared. The further characteristic is that the magnetization, the order parameter of the new state, rises as a power law $(T_c - T)^\beta$ where β is the critical exponent. An added fact is that the basic magnetic moment $\mu = iA$ defined as an area A times a loop current $i = dq/dt$ is related to an angular momentum L by the gyromagnetic ratio $e/2m$. Thus, we have $\mu = (e/2m) L$.

An important special case, uniaxial anti-ferromagnetism, turns out to be closely related to the hurricane transition, through its connection to the "Ising model." An antiferromagnet is a system where two opposite and unequal moments appear in each unit cell. A uniaxial system is one in which the magnetization is directed only along a particular direction, the z-direction. The Ising model is a linear array of equal moments, with the nearest neighbor interaction, which each can be directed in either the plus or minus z-direction. It has been well established (see, for example, the book of Nigel Goldenfeld) that the uniaxial antiferromagnet $DyAlO_3$, with critical exponent 0.311, is an experimental realization of the Ising model. Following Goldenfeld, one might take that exponent as the experimental value of the Ising model exponent. However, an authoritative review has given a theoretical value of the Ising model exponent as 0.3265 (Polisetto and Vicari, 2002). They provided 21 separate experimental values of the exponent β ranging from 0.111 to 0.341, and whose average value is 0.3197. Taking all these more recent values into account, we suggest the Ising model universality class critical exponent as 0.323, and that this number is in agreement with the various errors involved with the nominal observed value of $1/3$. The experts, Polisetto and Vicari, make the statement that the renormalization group approach of K. G. Wilson "explains...why fluids and uniaxial antiferromagnets behave in an identical way at the critical point," certainly to imply the exponents are identical. So, there is a possibility that the value 0.323, which we attribute to the universality class, is the preferred value. In that case, the critical exponent for the wind speed is 0.323 and that for the PDI is 0.969. These values are consistent with the inexact available data for PDI vs temperature shown in Figs. 1.5 and 1.6.

The Ising model behavior is illustrated in Fig. 5.1, where the measured quantity, proportional to the magnetization order parameter, is the magnetoelectric effect α. In the figure, the ordinate is the magnetoelectric constant at temperature T, normalized by its value at 1.4 K, well below the transition temperature, called the Neel temperature T_N = 3.525 K for the antiferromagnet $DyAlO_3$. As explained in detail by Nigel Goldenfeld (1992), this antiferromagnetic system is reliable as representing typical behavior near a ferromagnetic transition. Further, the exponent found, 0.311, should not only apply to all uniaxial antiferromagnets but also to a set of other systems.

5.1.2 Critical Exponents and Universality Class

Hurricanes are unrelated to magnetism but in the same "universality class." The value of the critical exponent in each universality class can be arrived at by using the "renormalization group". This generality is surprising and led to the Nobel Prize. The consequences of this relatively new body of knowledge for rainfall and hurricanes are only beginning to be explored.

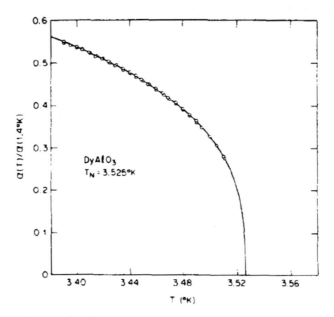

Figure 5.1 Power-law fit to magnetic transition showing with a precision that the critical exponent is 0.311, a general result within the Ising model universality class. Reproduced from Holmes et al., 1971. Copyright (1971), with permission from Elsevier.

The main result is that the order parameter for the exemplary antiferromagnet Dy AlO_3 is shown by Holmes et al., to vary as const $(T_N - T)^{0.311}$, for high accuracy, near the transition point. The high accuracy of the fit is seen in Fig. 5.1, extending to 3.39 K from 3.525 K. This range, 0.135 K, is 0.038 of T_N. The equivalent temperature range, where the power law is extremely accurate, for the hurricane case, with T_c = 26.5 °C = 299.5 K, would then be 11.4 K, extending the expected range of accuracy of the power-law fit (see Fig. 1.6) to 37.9 °C or 112.3 °F. Universality has been invoked for the following reasons. First, Wolf (2020) showed that the exponent for the PDI is a nominal unity. Since the PDI scales as the cube of the velocity, the more fundamental exponent for the velocity, is nominally 0.333, close to that for the Ising model (Fig. 5.1). Second, one can argue that the hurricane is in the same universality class as the magnetic Ising model because one can essentially map the Hamiltonian of the hurricane onto that of the Ising model by multiplying by the gyromagnetic ratio $e/2m$. The magnet is a vertically directed moment and the hurricane via the gyromagnetic ratio is a vertically directed angular momentum vector $L = Mvr$. If the two cases are in the same universality class, we can think of rewriting the ordinate scale of Fig. 5.1 in meters per second and changing the temperature scale to run from 40 °C to the transition at 26.5 °C. The basic physics allows one to state that the precision seen in the antiferromagnetic fit will be expected for the hurricane case, which is of course impossible to accurately measure since one cannot create hurricanes in the laboratory, while watching the thermometer. If this is correct, one can be confident that the power-law dependence is adequate for any reasonable scenario of global warming, avoiding sea temperatures of 112 °F. This view relegates the extensive potential intensity theory to possibly describing a high-temperature asymptotic region that, however, is unlikely to be reached in nature.

5.2   Relaxational Effects, Including Rainfall

In addition to phases of matter, a second class of phenomena, including tropical and eyewall cloud rainfall, which we may call relaxational effects, exhibit critical behavior. These phenomena include avalanches, earthquakes, and rainfall. In each case, stress builds up, called the tuning parameter, leading to a statistical release of the stress in the onset of the new phenomenon, the release of the snow in the avalanche, or the "quaking" of the earth or the onset of precipitation P.

For precipitation, the stress parameter is w, the total precipitable water, and the integral of water as vapor is in a vertical column, available in satellite data. This is a measure of humidity and is the parameter that was measured to form the image in Fig. 1.3.

Critical phenomena have been closely identified by Peters and Neelin (2006) in precipitation patterns above the tropical oceans. The rainfall precipitation P in mm/h is shown in Fig. 5.2 to rise according to a characteristic power law:

$$P = a \, (w - w_c)^{\beta}, \qquad\qquad (5.1)$$

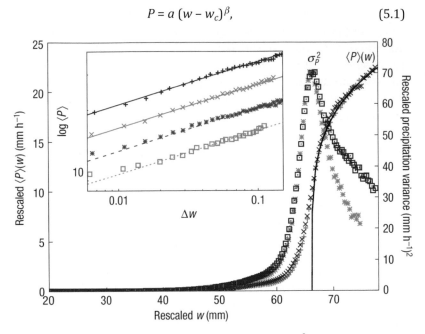

Figure 5.2 Order parameter P and susceptibility σ^2. The consolidated precipitation rates P (left scale, mm/h) and their variances σ^2 (right scale, mm^2/h^2) for eastern Pacific (red), western Pacific (green), Atlantic (blue), and Indian Ocean (pink), as well as power-law fit above the critical point (solid line) vs rescaled water content w in mm. These curves are a consolidation of curves as shown in Fig. 5.3. The upper-left inset shows on double-logarithmic scales, the precipitation rate as a function of the reduced water vapor. The straight lines in the inset all have a slope of 0.215, identified as a universal exponent. Reprinted from Peters and Neelin, 2006, with permission of Springer Nature.

with precipitation P rising sharply above a critical value of w, the local vertical atmospheric water vapor content, at about $w_c = 66$ mm of equivalent liquid water. Here β is the critical exponent and a is arbitrary. In careful studies of satellite data from the four principal tropical basins, namely, the Indian Ocean, the Atlantic, the western Pacific, and the eastern Pacific, the characteristic exponent was measured by Peters and Neelin to be universal at a value of 0.215. The precipitation at the midpoint of the sharp rise shown in Fig. 5.2 is about 12 mm/h, or about 0.5 inch/h. The order parameter, the rainfall rate, as a function of the column content of water vapor describes, above the critical value, a regime of turbulent convection and rainfall, as occurs in a thunderstorm. This rainfall is a value near that of the average rainfall under a tropical cyclone, known in the Atlantic as a hurricane.

Another confirmed expectation from the theory of phase transitions that is shown in these data on the right scale, is a peak in the precipitation variance σ^2 near w_c. In simple terms, the variance of a set of numbers is the average of the squares of the differences between the individual numbers and their mean value. The square root of this average is called the standard deviation. These properties are related to long-range spatial correlations in the critical region.

The temperature dependence involved in the rainfall data is made clearer in Fig. 5.3 after Neelin et al., 2008.

Rethinking convective quasi-equilibrium 2591

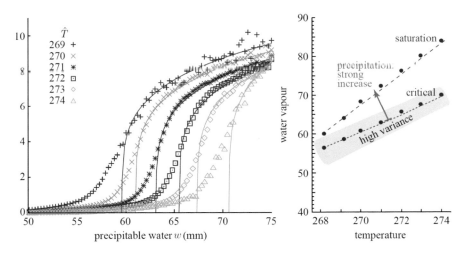

Figure 5.3 Left panel: Rising increments of average precipitation P in temperature bins (one Kelvin) vs column precipitable water in mm. (Satellite data combined from tropical western Pacific, eastern Pacific, and the Atlantic). Onsets of precipitation are seen in the accurate power-law fitting curves (Eq. (5.1)) to increase regularly with average temperature, as shown in Fig. 5.2. Right panel: The critical value w_c of column water vapor at which the transition to strong convection occurs vs. vertically averaged tropospheric temperature. It also showed saturation values of vertically integrated water vapor. Reproduced from Neelin et al., 2008. Copyright © 2008 The Royal Society.

5.2.1 Rainfall in a Hurricane

In their study of rainfall behavior as a continuous phase transition, Peters and Neelin (2009) have included rainfall data from Hurricane Katrina of 2005.

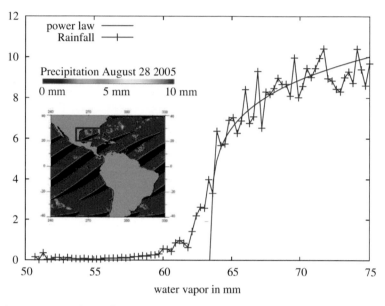

Figure 5.4 Conditionally averaged rain rate in a sample of satellite pixels containing measurements of Hurricane Katrina, including all temperatures: Gulf of Mexico (red box, 80 W to 100 W; 20 N to 30 N, August–September 2005). Red crosses are data, and the blue line is based on Eq. (5.1) using β = 0.215. Reproduced from Peters and Neelin, 2009, with permission of World Scientific Publishing.

In Fig. 5.4, the shown data points are that most of the strong precipitation occurred inside a hurricane (Hurricane Katrina of 2005). The authors did not stratify by temperature as was done in the earlier figures because the statistics of the data set were poorer. Due to the relatively frequent occurrence of strong rainfall and high-water vapor in the sample, it is nonetheless possible to observe a dependence similar to Eq. (5.1). The hurricane, by this measure, does not look fundamentally different from deep convective rain from less organized weather systems.

Chapter 6

Summary: Hurricanes as a Phase of Matter

6.1   Transition Temperature

One of the usual features of a phase change is a transition temperature. The literature on hurricanes has long recognized that hurricanes do not form when the sea surface temperature is below 26.5 °C (Palmen, 1948). A recent careful compendium of hurricane formation temperatures is that of Dare and McBride (2011).

The sharply rising curve in Fig. 6.1 is certainly of the type that would be interpreted as a phase transition, for example, a superconductive transition or a ferromagnetic transition.

As explained in Chapter 5, the additional and stronger reason to regard hurricane formation as a phase transition is the temperature dependence of the characteristic feature, the azimuthal wind speed V, shown in the work of Emanuel (2005) by Wolf (2020) to be $V = $ const. $(T - T_c)^{1/3}$. This behavior is inferred from the directly observed PDI = const. $(T - T_c)$ and the definition of PDI that is proportional to the cube of the wind speed. The original papers of Emanuel (2005 and 2007) show accurate fits using this algorithm for hurricanes in the Pacific as well as the Atlantic cyclone basins. The same T_c value in each

Physics and Future of Hurricanes
Edward L. Wolf
Copyright © 2023 Jenny Stanford Publishing Pte. Ltd.
ISBN 978-981-4968-54-6 (Hardcover), 978-1-003-33125-4 (eBook)
www.jennystanford.com

of these plots was found applicable to many hurricanes over decades. The T_c values were neither recognized nor disclosed in the original papers but are close to 26.5 °C, in agreement with the work of Dare and McBride (2011).

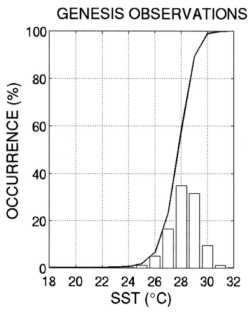

Figure 6.1 Number of hurricanes vs sea surface temperature from a study of several thousand hurricanes worldwide. The behavior is described in the literature as a threshold temperature near 26.5 °C. Following Wolf (2020) as shown in Fig. 1.6, we believe the correct description is a transition temperature to a new phase of matter. Reproduced from Dare and McBride, 2011. © American Meteorological Society. Used with permission.

6.2 A Distinct Organization of Matter in Ising Universality Class

The precise circular eye and eyewall showed in Fig. 1.3 cannot be imagined other than as representing a distinct detailed organization of matter quite unlike a collection of thunderstorms. The direct infrared camera image shown for Hurricane Harvey at its landfall reveals a system of a different kind, including a precisely dimensioned circular eyewall just outside the empty eye region, all unperturbed by the passage of the hurricane over the coastline.

This image alone reveals the hurricane as a distinct phase of matter. The detailed circular symmetry of the inner region of the hurricane shown here is in agreement with photographs taken from research aircraft flown inside the eye. In such photographs, the eyewall cloud is seen to be accurately circular on a large scale, and rising in a martini-glass shape diagonally upward and outward for 10 km or more. The smoothness of these features is evidence of a new organization in the new phase of matter.

The theory of the second-order phase transitions finds that phases of matter are organized in universality classes that share identical critical exponents in spite of wide dissimilarity. For example, it is known that the magnetization vs temperature of uniaxial antiferromagnets, such as Dy AlO_3, as shown in Fig. 5.1, shares precisely its critical exponent with that of the liquid-vapor transition of simple fluids. This is a result of the renormalization group theory of K. Wilson, 1983. Both of these systems are known to be described by the Ising model, which we can view as an idealized uniaxial antiferromagnet. This Ising Model exponent we have discussed above is 0.323±0.01. This value within experimental error is the same as the 1/3 implied by Fig. 1.6, making use of the fact that the PDI scales as the cube of the wind speed V.

We assert that the hurricane phase is in the Ising universality class, first, because the critical exponent is correct for that assignment. Further, the nature of the hurricane phase is a vertically directed angular momentum, generated by the azimuthal wind flow. This resembles the vertically directed magnetic moment of the uniaxial antiferromagnet whose nature, again, is a magnetic moment directed vertically and related to a vertical uniaxial angular momentum by the universal gyromagnetic ratio $e/2m$.

Having made this identification, we realize that the Ising model exponent offers a more accurate value for the hurricane phase exponent, admitting that the exponent for the PDI is a nominal unity from Fig. 1.6 and Emanuel's fitting algorithm (Eq. (1.17)) but is not determined to better than 10%. Thus, we conclude that the hurricane velocity exponent is 0.323±0.01 and that the exponent for hurricane power, the PDI, is 0.969±0.03. These slightly improved values are within experimental error of the originally published values, 1/3 and unity (Wolf, 2020).

6.3 Precise Prediction of Hurricane Power vs Ocean Temperature

Following the discussion above, a precise prediction of the PDI vs ocean temperature is:

$$\text{PDI} = a \, (T - T_c)^{0.969 \pm 0.03}, \tag{6.1}$$

where T_c = 26.5±0.01 °C and a is arbitrary. The precise value is based on the critical exponent for the Ising model, but, with the error range, overlaps the simple statement of linearity following from Fig. 1.6 and also overlaps the value that follows from the L. M. Holmes data of Fig. 5.1.

Correspondingly, the precise value for the maximum wind speed is:

$$V = b \, (T - T_c)^{0.323 \pm 0.01}, \tag{6.2}$$

where b is arbitrary. The analysis given in conjunction with Fig. 5.1 suggests that the formulas should be accurate for sea surface temperatures as high as 38 °C, a temperature range that includes any likely climatic future.

The replotted PDI data, shown in Fig. 1.6, are consistent with, confirm Eqs. (6.1) and (6.2), in the straight line shown extrapolating to the critical temperature of 26.5 °C, a value in close agreement with Dare and McBride. This line in Fig. 1.6 increases from PDI near 1.5 (PDI in units 10^{11} m^3 s^{-2}) at 26.85 °C to about 4 at 27.6 °C. This is a rate of change in PDI of 3.33 per degree. If we extrapolate 2 degrees higher, say by the year 2100, from the value 4 near 2005, we find PDI = 10.66, which is a factor 10.66/4 = 2.67 larger. So, the PDI roughly triples if there is a 2 °C increase in sea temperature. In terms of wind speed, the increase will be proportional to the cube root, thus an increase in wind speed by a factor $(2.67)^{1/3}$ = 1.386. So, the maximum wind scale would be increased by 39% while the expected monetary losses, scaling with power expended would nearly triple, increasing by a factor of 2.67.

6.4 Conclusion

In conclusion, we confirm an alternative view (Wolf, 2020) of extensive and closely fitted data sets (Emanuel, 2005, 2007) of the hurricane PDI vs measured sea surface temperature T. These data provide excellent detailed fits if the measured temperatures are subjected to offsets that are equivalent to choosing a critical temperature for a tropical cyclone, via fitting law (Eq. (1.17)), with a linear dependence on $(T - T_c)$. The inferred Eq. (1.17) is confirmed by direct data replotted in Fig. 1.6. It is suggested that the data and fits are strong evidence that the appearance of hurricanes can be viewed as a critical point, of the macroscopic warm ocean-atmosphere system. This also reveals a stronger temperature dependence of the cyclone strength, represented by the PDI, which seems of interest and concern from the point of view of rising ocean temperature. We here find that precise values of the critical exponents for the PDI and maximum wind speed, respectively, are 0.969±0.03 and 0.323±0.01. These slight improvements on our earlier values (Wolf 2020, 2021) are found by recognizing the hurricane phase transition to be in the Ising model universality class, known also to describe the transition of the uniaxial antiferromagnet (Fig. 5.1) and the liquid-vapor transition of simple fluids.

References

Bister, M., and Emanuel, K. (1998), "Dissipative heating and hurricane intensity," *Meteor. Atmos. Phys.*, 65, 233.

Bister, M., and Emanuel, K. A. (2002), "Low frequency variability of tropical cyclone potential intensity I, Interannual to interdecadal variability," *J. Geophys. Res.*, 107, 4801.

Bluestein, H., McCaul, E., Byrd, G., and Woodall, G. (1988), "Mobile sounding observation of a tornadic storm near the Dry line, The Canadian, Texas storm of 7 May 1986," *Mon. Weather Rev.*, 116, 1790.

Cappucci, M. (2020b), "Iowa is as prone to destructive Derechos as Florida is to hurricanes," *Washington Post*, August 19, 2020. Cappucci, M. (2020c), "Saharan dust is suppressing hurricane activity over the Atlantic, Don't count on it staying that way," *The Washington Post*, June 23, 2020.

Dare, R., and McBride, J. (2011), "The threshold sea surface temperature condition for tropical cyclones," *J. Climate*, 24, 4570.

Davis, R., and Paxton, C. (2005), "How swells of hurricane Isabel impacted Southeast Florida," *Bull. Am. Meteor Soc.*, 86, 1065.

Emanuel, K. (1986), "An air-sea interaction theory for tropical cyclones, Part I: Steady-state maintenance," *J. Atmospheric Sci.*, 93, 585.

Emanuel, K. (1991), "The theory of hurricanes," *Annu. Rev. Fluid Mech.*, 23, 179.

Emanuel, K. (1999), "The power of a hurricane: An example of reckless driving on the information superhighway," *Weather*, 54 (4), 107.

Emanuel, K. (2005), *The Divine Wind: The History and Science of Hurricanes*, Oxford University Press.

Emanuel, K. (2005a), "Increasing destructiveness of tropical cyclones over the past 30 years," *Nature*, 436, 686.

Emanuel, K. (2007), "Environmental factors affecting tropical cyclone power dissipation," *J. Climate*, 20, 5497.

Emanuel, K. (2018), "100 Years of Progress in Tropical Cyclone Research," Chapter 15 in *AMS Monographs*, Vol. 59, American Meteorological Society.

Emanuel, K., and Rotunno, R. (2011), "Self-stratification of tropical cyclone outflow, Part I: Implications for storm structure," *J. Atmospheric Sci.*, 68, 2236.

Goldenfeld, N. (1992), *Lectures on Phase Transitions and the Renormalization Group* Frontiers in Physics, CRC Press.

Hallett, J., and Mossop, S. (1974), "Production of secondary ice particles during the riming process," *Nature*, 249, 26.

Heymsfield, A., and Mossop, S. (1984), "Temperature dependence of secondary ice production during soft hail production by riming," *Q. J. R. Meteorol. Soc.*, 71, 4500.

Heymsfield, A., Szakall, M., Jost, A., and Giammanco, I. (2018), "A comprehensive observational study of graupel and hail terminal velocity mass flux and kinetic energy," *J. Atmospheric Sci.*, 75, 3861.

Holmes, L. M., Van Uitert, L. G., and Hull, G. W. (1971), "Magnetoelectric effect and critical behavior in the Ising-like antiferromagnet $DyAlO_3$," *Solid State Commun.*, 9, 1373.

Houze, R., Jr. (2010), "Clouds in tropical cyclones," *Mon. Weather Rev.*, 138, 293.

Irvine, P., Emanuel, K., He, J., Horowitz, L., Vecchi, G., and Keith, D. (2019), "Halving warming with idealized solar geoengineering moderates key climate hazards," *Nat. Climate Change*, 9, 295.

Liu, W., Tang, W., and Niiler, P. (1991), "Humidity profiles over the ocean," *J. Climate*, 4, 1023.

Lstiburek, J. (2014), "How buildings stack up," *ASRAE J.*, 56, 42.

Lu, P., Lin, N., Emanuel, K., Chavas, D., and Smith, J. (2018), "Assessing hurricane rainfall mechanisms using a physics-based model: hurricanes Isabel (2003) and Irene (2011)," *J. Atmospheric Sci.*, 75, 2337.

Murnane, R., and Liu, K.-B. (eds.) (2004), *Hurricanes and Typhoons: Past, Present and Future*, Columbia University Press, New York.

Neelin, J., Peters, O., Lin, J., Hales, K., and Holloway, C. (2008), "Rethinking convective quasi-equilibrium: Observational constraints for stochastic convective schemes in climate models," *Phil. Trans. Royal Soc. A*, 366, 2581.

Palmen, E. (1948), "On the formation and structure of tropical hurricanes," *Geophysica*, 3, 26.

Peters, O. M., and Neelin, J. D. (2006), "Critical phenomena in atmospheric precipitation," *Nature Phys.*, 2, 393.

Peters, O., and Neelin, J. (2009), "Atmospheric convection as a continuous phase transition: Further evidence," *Int. J. Mod. Phys. B.*, 23, 5453.

Polisetto, A., and Vicari, E. (2002), "Critical phenomena and renormalization group theory," *Phys. Rep.*, 368, 547.

Randall, D. (2012), *Atmosphere, Clouds and Climate*, Princeton University Press, Princeton, p. 67.

Smith, R., Montgomery, M., and Van Sang, N. (2009), "Tropical cyclone spin-up revisited," *Q. J. R. Meteorol. Soc.*, 135, 1321.

Smith, R., and Montgomery, M. (2016), "Understanding hurricanes," *Weather*, 71, 219.

Wilson, K. G. (1983), "The renormalization group and critical phenomena," *Rev. Mod. Phys.*, 55, 583.

Wolf, E. L. (2020), "Critical behavior of tropical cyclones," *Theor. Appl. Climatol.*, 139(3), 1231.

Wolf, E. L. (2021), "Precise prediction of hurricane power vs ocean temperature," *Int. J. Atmospheric Oceanic Science*, 5(1), 1–5.

Zou, X., and Wu, Y. (2005), "On the relationship between total ozone mapping spectrometer (TOMS) and hurricanes," *J. Geophys. Res.*, 110, D06109.3

Index